P9-DVG-528

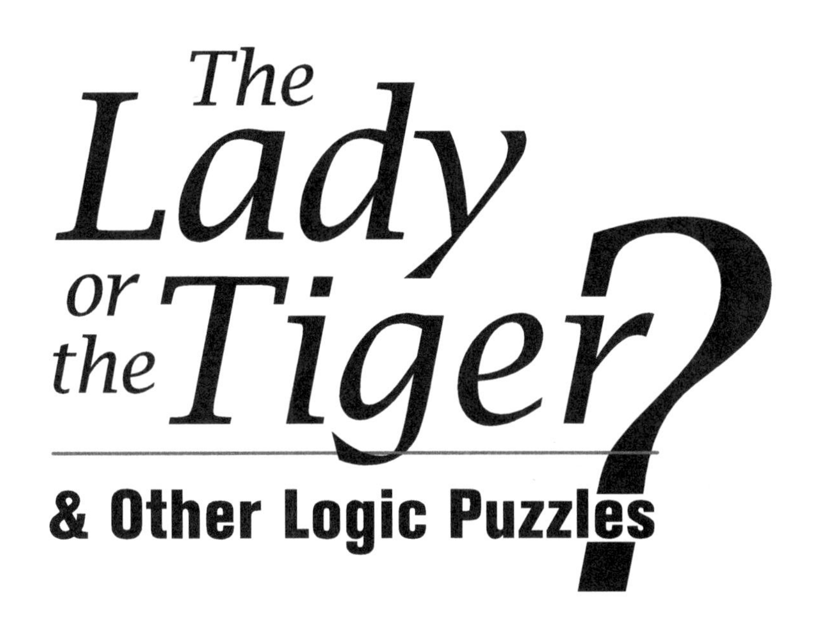

Raymond M. Smullyan

DOVER PUBLICATIONS, INC.
Mineola, New York

Bibliographical Note

This Dover edition, first published in 2009, is an unabridged republication of the work originally published by Times Books (Random House), New York, in 1982.

Library of Congress Cataloging-in-Publication Data

Smullyan, Raymond M.
The lady or the tiger? and other logic puzzles / Raymond Smullyan — Dover ed.
p. cm.
Originally published: New York : Times Books, c1982.
ISBN-13: 978-0-486-47027-6 (pbk.)
ISBN-10: 0-486-47027-X (pbk.)
1. Logic puzzles. 2. Mathematical recreations. 3. Philosophical recreations. I. Title.

GV1493.S626 2009
793.73—dc22

2008050048

Manufactured in the United States by Courier Corporation
47027X03
www.doverpublications.com

Contents

CONTENTS

Preface

Among the numerous fascinating letters I have received concerning my first puzzle book (whose name I can never remember!), one was from the ten-year-old son of a famous mathematician who was a former classmate of mine. The letter contained a beautiful original puzzle, inspired by some of the puzzles in my book which the boy had been avidly reading. I promptly phoned the father to congratulate him on his son's cleverness. Before he called the boy to the phone, the father said to me in soft conspiratorial tones: "He is reading your book and loves it! But when you speak to him, don't let him know that what he is doing is math, because he *hates* math! If he had any idea that this is really math, he would stop reading the book immediately!"

I mention this because it illustrates a most curious, yet common, phenomenon: So many people I have met claim to hate math, and yet are enormously intrigued by any logic or math problem I give them, provided I present it in the form of a puzzle. I would not be at all surprised if good puzzle books prove to be one of the best cures for so-called "math anxiety." Moreover, any math treatise *could* be written in the format of a puzzle book! I have sometimes wondered what would have happened if Euclid had written his classic *Ele-*

ments in such a format. For example, instead of stating as a theorem that the base angles of an isosceles triangle are equal, and then giving the proof, he could have written: "*Problem:* Given a triangle with two equal sides, are two of the angles necessarily equal? Why, or why not? (For the solution, see page —.)" And similarly with all the rest of his theorems. Such a book might well have become one of the most popular puzzle books in history!

In general, my own puzzle books tend to be different from others in that I am primarily concerned with puzzles that bear a significant relation to deep and important results in logic and mathematics. Thus, the real aim of my first logic book was to give the general public an inkling of what Gödel's great theorem was about. The volume you are now holding goes still further in this direction. I used the manuscript of it in a course entitled "Puzzles and Paradoxes," where one of the students remarked to me: "You know, this whole book—particularly parts Three and Four—has much the flavor of a mathematical novel. I have never before seen anything like it!"

I think the phrase "mathematical novel" is particularly apt. Most of this book is indeed in the form of a narrative, and a good alternative title for it would be "The Mystery of the Monte Carlo Lock," since the last half concerns a case in which Inspector Craig of Scotland Yard must discover a combination that will open the lock of a safe in Monte Carlo to prevent a disaster. When his initial efforts to crack the case prove unsuccessful, the inspector returns to London, where he serendipitously renews acquaintance with a brilliant and eccentric inventor of number machines. Together they team up with a mathematical logician, and soon the three find themselves in ever-deepening waters that flow into the very heart of Gödel's great discovery. The Monte Carlo lock, of course, turns out to be a "Gödelian" lock in disguise, its *modus operandi* beautifully reflecting a fundamental idea of Gödel's that has basic ramifications in many scientific theories dealing with the remarkable phenomenon of self-reproduction.

As a noteworthy dividend, the investigations of Craig and his friends lead to some startling mathematical discoveries hitherto unknown to either the general public or the scientific community. These are "Craig's laws" and "Fergusson's laws," which are published here for the very first time. They should prove of equal interest to the layman, the logician, the linguist, and the computer scientist.

This whole book has been great fun to write, and should be equal fun to read. I am planning several sequels. Again I wish to thank my editor, Ann Close, and the production editor, Melvin Rosenthal, for the wonderful help they have given me.

RAYMOND SMULLYAN

Elka Park, N.Y.
February 1982

PART ONE

THE LADY OR THE TIGER ?

◆ 1 ◆

Chestnuts—Old and New

I would like to begin with a series of miscellaneous arithmetical and logical puzzles—some new, some old.

1 ◆ How Much?

Suppose you and I have the same amount of money. How much must I give you so that you have ten dollars more than I? (Solutions come at the end of each chapter.)

2 ◆ The Politician Puzzle

A certain convention numbered one hundred politicians. Each politician was either crooked or honest. We are given the following two facts:

(1) At least one of the politicians was honest.

(2) Given any two of the politicians, at least one of the two was crooked.

Can it be determined from these two facts how many of the politicians were honest and how many were crooked?

3 ◆ Old Wine in a Not-so-new Bottle

A bottle of wine cost ten dollars. The wine was worth nine dollars more than the bottle. How much was the bottle worth?

4 ◆ How Much Profit?

The amazing thing about this puzzle is that people always seem to fight over the answer! Yes, different people work it out in different ways and come up with different answers, and each insists his answer is correct. The puzzle is this:

A dealer bought an article for $7, sold it for $8, bought it back for $9, and sold it for $10. How much profit did he make?

5 ◆ Problem of the Ten Pets

The instructive thing about this puzzle is that although it can easily be solved by using elementary algebra, it can also be solved without any algebra at all—just by plain common sense. Moreover, the common-sense solution is, in my judgment, far more interesting and informative—and certainly more creative—than the algebraic solution.

Fifty-six biscuits are to be fed to ten pets; each pet is either a cat or a dog. Each dog is to get six biscuits, and each cat is to get five. How many dogs and how many cats are there?

Any reader familiar with algebra can get this immediately. Also, the problem can be solved routinely by trial and error: there are eleven possibilities for the number of cats (anywhere from zero to ten), so each possibility can be tried until the correct answer is found. But if you look at this problem in just the right light, there is a surprisingly simple solution that involves neither algebra nor trial-and-error. So, I urge even those of you who have gotten the answer by your own method to consult the solution I give.

6 • Large Birds and Small Birds

Here is another puzzle that can be solved either by algebra or by common sense, and again I prefer the common-sense solution.

A certain pet shop sells large birds and small birds; each large bird fetches twice the price of a small one. A lady came in and purchased five large birds and three small ones. If, instead, she had bought three large birds and five small birds, she would have spent $20 less. What is the price of each bird?

7 • The Disadvantages of Being Absent-minded

The following story happens to be true:

It is well known that in any group of at least 23 people, the odds are greater than 50 percent that at least two of them will have the same birthday. Now, I was once teaching an undergraduate mathematics class at Princeton, and we were doing a little elementary probability theory. I explained to the class that with 30 people instead of 23, the odds would become enormously high that at least two of them had the same birthday.

"Now," I continued, "since there are only nineteen students in this class, the odds are much *less* than fifty percent that any two of you have the same birthday."

At this point one of the students raised his hand and said,

"I'll bet you that at least two of us here have the same birthday."

"It wouldn't be right for me to take the bet," I replied, "because the probabilities would be highly in my favor."

"I don't care," said the student, "I'll bet you anyhow!"

"All right," I said, thinking to teach the student a good lesson. I then proceeded to call on the students one by one to announce their birthdays until, about halfway through, both I and the class burst out laughing at my stupidity.

The boy who had so confidently made the bet did not know the birthday of anyone present except his own. Can you guess why he was so confident?

8 • Republicans and Democrats

In a certain lodge, each member was either a Republican or a Democrat. One day one of the Democrats decided to become a Republican, and after this happened, there was the same number of Republicans as Democrats. A few weeks later, the new Republican decided to become a Democrat again, and so things were back to normal. Then another Republican decided to become a Democrat, at which point there were twice as many Democrats as Republicans.

How many members did the lodge contain?

9 • A New "Colored Hats" Problem

Three subjects—A, B, and C—were all perfect logicians. Each could instantly deduce all consequences of any set of premises. Also, each was aware that each of the others was a perfect logician. The three were shown seven stamps: two red ones, two yellow ones, and three green ones. They were then blindfolded, and a stamp was pasted on each of their foreheads; the remaining four stamps were placed in a drawer. When the blindfolds were removed, A was asked, "Do you know one color that you definitely do not have?" A replied, "No." Then B was asked the same question and replied, "No."

Is it possible, from this information, to deduce the color of A's stamp, or of B's, or of C's?

10 • For Those Who Know the Rules of Chess

I would like to call your attention to a fascinating variety of chess problem which, unlike the conventional chess problem—White to play and mate in so-many moves—involves an analysis of the past history of the game: how the position arose.

Inspector Craig of Scotland Yard,* whose interest in this type of problem was equal to that of Sherlock Holmes,† once walked with a friend into a chess club, where they came across an abandoned chessboard.

"Whoever played this game," said the friend, "certainly doesn't know the rules of chess. The position is quite impossible!"

"Why?" asked Craig.

"Because," replied the friend, "Black is now in check from both the White rook and the White bishop. How could White possibly have administered this check? If he has just

* Inspector Craig is a character from my previous book of logic puzzles, *What Is the Name of This Book?*

† My book *The Chess Mysteries of Sherlock Holmes* contains many puzzles of this genre.

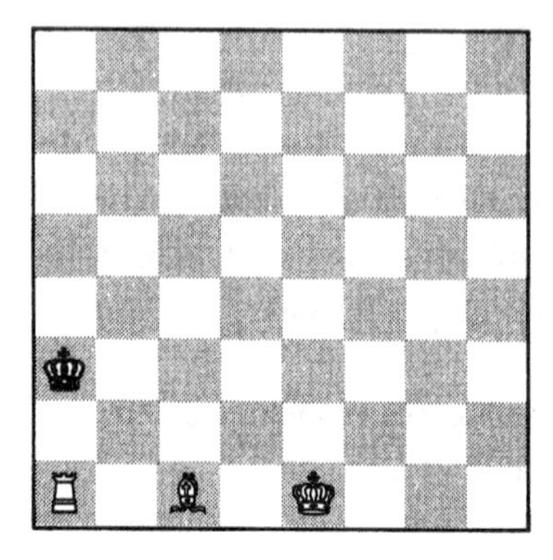

moved the rook, the Black king would already be in check from the bishop, and if he has just moved the bishop, the king would already be in check from the rook. So, you see, the position is impossible."

Craig studied the position for a while. "Not so," he said, "the position, though exceedingly bizarre, is well within the bounds of legal possibilities."

Craig was absolutely right! Despite all appearances to the contrary, the position really is possible, and it can be deduced what White's last move was. What was it?

• SOLUTIONS •

1 • A common wrong answer is $10. Now, suppose we each had, say, $50. If I gave you $10, you would then have $60 and I would have only $40; hence you would have $20 more than I, rather than $10.

The correct answer is $5.

2 • A fairly common answer is "50 honest and 50 crooked." Another rather frequent one is "51 honest and 49 crooked."

Both answers are wrong! Now let us see how to find the correct solution.

We are given that at least one person is honest. Let us pick out any one honest person, whose name, say, is Frank. Now pick any of the remaining 99; call him John. By the second given condition, at least one of the two men—Frank, John—is crooked. Since Frank is not crooked, it must be John. Since John arbitrarily represents any of the remaining 99 men, then each of those 99 men must be crooked. So the answer is that one is honest and 99 are crooked.

Another way of proving it is this: The statement that given any two, at least one is crooked, says nothing more nor less than that given any two, they are not both honest; in other words, no two are honest. This means that at most one is honest. Also (by the first condition), at least one is honest. Hence exactly one is honest.

Do you prefer one proof to the other?

3 • A common wrong answer is $1. Now, if the bottle were really worth a dollar, then the wine, being worth $9 more than the bottle, would be worth $10. Hence the wine and bottle together would be worth $11. The correct answer is that the bottle is worth 50¢ and the wine is worth $9.50. Then the two add up to $10.

4 • Some argue as follows: After having bought the article for $7 and sold it for $8, he has made a dollar profit. Then, by buying the article back for $9 after having sold it for $8, he loses a dollar; hence at this point he is even. But then (the argument continues) by selling for $10 what he has bought for $9, he has made a dollar again; therefore, his total profit is $1.

Another argument leads to the conclusion that the dealer broke even: When he sold the article for $8 after having bought it for $7, he made $1 profit. But then he loses $2 by buying back for $9 the item for which he originally paid $7, and so at this point he is $1 in the hole. Then he gets back the dollar by selling for $10 the article for which he last paid $9, and so now he is even.

Both arguments are wrong; the correct answer is that the dealer made $2. There are several ways to arrive at this—one such goes as follows: First, after selling for $8 the article for which he has paid $7, he has clearly made $1. Now, suppose that instead of buying back the *same* article for $9 and then selling it for $10, he were to buy a *different* article for $9 and sell it for $10. Would this really be any different from a purely economic point of view? Of course not! He would obviously, then, be making another dollar on the buying and selling of this second article. Thus, he has made $2.

Another and very simple proof: The dealer's total outlay is $7 + $9 = $16, and his total return is $8 + $10 = $18, giving a profit of $2.

For those not convinced by these arguments, let us suppose that the dealer has a certain amount of money—say, $100—at the opening of the day and that he makes just these four deals. How much will he have at the close of the day? Well, first he pays $7 for the article, leaving him with $93. Then he sells the article for $8, giving him $101. Next he buys the article back for $9, bringing him down to $92. Finally, he sells the article for $10 and thus winds up with $102. So he has started the day with $100 and ended it with $102. How, then, could his profit be anything other than $2?

5 ◆ The solution I have in mind is this: First feed five biscuits to each of the ten pets; this leaves six biscuits. Now, the cats have already had their portion! Therefore, the six remaining biscuits are for the dogs, and since each dog is to get

one more biscuit, there must be six dogs, and thus four cats.

Of course, we can check out the solution: Six dogs each getting six biscuits accounts for thirty-six biscuits. Four cats each getting five biscuits accounts for twenty biscuits. The total (36 + 20) is 56, as it should be.

6 ◆ Since each large bird is worth two small birds, then five large birds are worth ten small birds. Hence five large birds plus three small birds are worth thirteen small birds. On the other hand, three large birds plus five small birds are worth eleven small birds. So the difference between buying five large and three small birds or buying three large and five small birds is the same as the difference between buying thirteen small birds and buying eleven small birds, which is two small birds. We know that the difference is $20. So two small birds are worth $20, which means one small bird is worth $10.

Let us check: A small bird is worth $10, and a large bird $20. Therefore, the lady's bill for five large and three small birds was $130. Had she bought three large and five small birds, she would have spent $110, which is indeed $20 less.

7 ◆ At the time I accepted the bet from the student, I had totally forgotten that two of the other students, who always sat next to each other, were identical twins.

8 ◆ There were twelve members: seven Democrats and five Republicans.

9 ◆ The only one whose color can be determined is C. If C's stamp were red, then B would have known that his stamp was not red by reasoning: "If my stamp were also red, then A, seeing two red stamps, would know that his stamp is not red.

But A does not know that his stamp is not red. Therefore, my stamp cannot be red."

This proves that if C's stamp were red, then B would have known that his stamp was not red. But B did not know that his stamp was not red; therefore, C's stamp cannot be red.

The same argument, replacing the word *red* with *yellow*, shows that C's stamp cannot be yellow either. Therefore, C's stamp must be green.

10 ◆ It is not given which side of the board is White and which side is Black. It may well *appear* that White is moving up the board, but if he really were, then the position *would* be impossible! The truth is that White must be moving down the board, and that before the last move, the position was this:

The circle on the lower left-hand square represents some Black piece (either a queen, rook, bishop, or knight; there is no way to know which). The White pawn then captures this Black piece and promotes to a rook, bringing the game to the present position.

Of course, one might well ask: "Why did White promote to a rook instead of a queen; is this not highly improbable?" The answer is that indeed it is highly *improbable,* but that any other last move is not merely improbable but impossible, and as Sherlock Holmes so wisely said to Watson: "When we have eliminated the impossible, whatever remains, however improbable, must be the truth."

◆ 2 ◆

Ladies or Tigers?

Many of you are familiar with Frank Stockton's story "The Lady or the Tiger?," in which the prisoner must choose between two rooms, one of which contains a lady and the other a tiger. If he chooses the former, he marries the lady; if he chooses the latter, he (probably) gets eaten by the tiger.

The king of a certain land had also read the story, and it gave him an idea. "Just the perfect way to try my prisoners!" he said one day to his minister. "Only, I won't leave it to chance; I'll have signs on the doors of the rooms, and in each case I'll tell the prisoner certain facts about the signs. If the prisoner is clever and can reason logically, he'll save his life—and win a nice bride to boot!"

"Excellent idea!" said the minister.

THE TRIALS OF THE FIRST DAY

On the first day, there were three trials. In all three, the king explained to the prisoner that each of the two rooms contained either a lady or a tiger, but it *could* be that there were tigers in both rooms, or ladies in both rooms, or then again, maybe one room contained a lady and the other room a tiger.

1 ◆ The First Trial

"Suppose both rooms contain tigers," asked the prisoner. "What do I do then?"

"That's your hard luck!" replied the king.

"Suppose both rooms contain ladies?" asked the prisoner.

"Then, obviously, that's your good luck," replied the king. "Surely you could have guessed the answer to that!"

"Well, suppose one room contains a lady and the other a tiger, what happens then?" asked the prisoner.

"In that case, it makes quite a difference which room you choose, doesn't it?"

"How do I know which room to choose?" asked the prisoner.

The king pointed to the signs on the doors of the rooms:

I	II
IN THIS ROOM THERE IS A LADY, AND IN THE OTHER ROOM THERE IS A TIGER	IN ONE OF THESE ROOMS THERE IS A LADY, AND IN ONE OF THESE ROOMS THERE IS A TIGER

"Is it true, what the signs say?" asked the prisoner.

"One of them is true," replied the king, "but the other one is false."

If you were the prisoner, which door would you open (assuming, of course, that you preferred the lady to the tiger)?

2 • The Second Trial

And so, the first prisoner saved his life and made off with the lady. The signs on the doors were then changed, and new occupants for the rooms were selected accordingly. This time the signs read as follows:

I	II
AT LEAST ONE OF THESE ROOMS CONTAINS A LADY	A TIGER IS IN THE OTHER ROOM

"Are the statements on the signs true?" asked the second prisoner.

"They are either both true or both false," replied the king.

Which room should the prisoner pick?

3 • The Third Trial

In this trial, the king explained that, again, the signs were either both true or both false. Here are the signs:

I	II
EITHER A TIGER IS IN THIS ROOM OR A LADY IS IN THE OTHER ROOM	A LADY IS IN THE OTHER ROOM

Does the first room contain a lady or a tiger? What about the other room?

THE SECOND DAY

"Yesterday was a fiasco," said the king to his minister. "All three prisoners solved their puzzles! Well, we have five trials coming up today, and I think I'll make them a little tougher."

"Excellent idea!" said the minister.

Well, in each of the trials of this day, the king explained that in the lefthand room (Room I), if a lady is in it, then the sign on the door is true, but if a tiger is in it, the sign is false. In the righthand room (Room II), the situation is the opposite: a lady in the room means the sign on the door is false, and a tiger in the room means the sign is true. Again, it is possible that both rooms contain ladies or both rooms contain tigers, or that one room contains a lady and the other a tiger.

4 • The Fourth Trial

After the king explained the above rules to the prisoner, he pointed to the two signs:

I	II
BOTH ROOMS CONTAIN LADIES	BOTH ROOMS CONTAIN LADIES

Which room should the prisoner pick?

5 • The Fifth Trial

The same rules apply, and here are the signs:

I	II
AT LEAST ONE ROOM CONTAINS A LADY	THE OTHER ROOM CONTAINS A LADY

6 • The Sixth Trial

The king was particularly fond of this puzzle, and the next one too. Here are the signs:

I	II
IT MAKES NO DIFFERENCE WHICH ROOM YOU PICK	THERE IS A LADY IN THE OTHER ROOM

What should the prisoner do?

7 • The Seventh Trial

Here are the signs:

I	II
IT DOES MAKE A DIFFERENCE WHICH ROOM YOU PICK	YOU ARE BETTER OFF CHOOSING THE OTHER ROOM

What should the prisoner do?

8 ◆ The Eighth Trial

"There are no signs above the doors!" exclaimed the prisoner.

"Quite true," said the king. "The signs were just made, and I haven't had time to put them up yet."

"Then how do you expect me to choose?" demanded the prisoner.

"Well, here are the signs," replied the king.

THIS ROOM CONTAINS A TIGER

BOTH ROOMS CONTAIN TIGERS

"That's all well and good," said the prisoner anxiously, "but which sign goes on which door?"

The king thought awhile. "I needn't tell you," he said. "You can solve this problem without that information.

"Only remember, of course," he added, "that a lady in the lefthand room means the sign which should be on that door is true and a tiger in it means the sign should be false, and that the reverse is true for the righthand room."

What is the solution?

THE THIRD DAY

"Confound it!" said the king. "Again all the prisoners won! I think tomorrow I'll have *three* rooms instead of two; I'll put a lady in one room and a tiger in each of the other two rooms. Then we'll see how the prisoners fare!"

"Excellent idea!" replied the minister.

"Your conversation, though flattering, is just a bit on the repetitious side!" exclaimed the king.

"Excellently put!" replied the minister.

9 • The Ninth Trial

Well, on the third day, the king did as planned. He offered three rooms to choose from, and he explained to the prisoner that one room contained a lady and the other two contained tigers. Here are the three signs:

I	II	III
A TIGER IS IN THIS ROOM	A LADY IS IN THIS ROOM	A TIGER IS IN ROOM II

The king explained that at most one of the three signs was true. Which room contains the lady?

10 • The Tenth Trial

Again there was only one lady and two tigers. The king explained to the prisoner that the sign on the door of the room containing the lady was true, and that at least one of the other two signs was false.

Here are the signs:

I	II	III
A TIGER IS IN ROOM II	A TIGER IS IN THIS ROOM	A TIGER IS IN ROOM I

What should the prisoner do?

11 ◆ First, Second, and Third Choice

In this more whimsical trial, the king explained to the prisoner that one of the three rooms contained a lady, another a tiger, and the third room was empty. The sign on the door of the room containing the lady was true, the sign on the door of the room with the tiger was false, and the sign on the door of the empty room could be either true or false. Here are the signs:

I	II	III
ROOM III IS EMPTY	THE TIGER IS IN ROOM I	THIS ROOM IS EMPTY

Now, the prisoner happened to know the lady in question and wished to marry her. Therefore, although the empty room was preferable to the one with the tiger, his first choice was the room with the lady.

Which room contains the lady, and which room contains the tiger? If you can answer these two questions, you should have little difficulty in also determining which room is empty.

THE FOURTH DAY

"Horrible!" said the king. "It seems I can't make my puzzles hard enough to trap these fellows! Well, we have only one more trial to go, but this time I'll *really* give the prisoner a run for his money!"

12 ♦ A Logical Labyrinth

Well, the king was as good as his word. Instead of having three rooms for the prisoner to choose from, he gave him nine! As he explained, only one room contained a lady; each of the other eight either contained a tiger or was empty. And, the king added, the sign on the door of the room containing the lady is true; the signs on doors of all rooms containing tigers are false; and the signs on doors of empty rooms can be either true or false.

Here are the signs:

I	II	III
THE LADY IS IN AN ODD-NUMBERED ROOM	THIS ROOM IS EMPTY	EITHER SIGN V IS RIGHT OR SIGN VII IS WRONG

IV	V	VI
SIGN I IS WRONG	EITHER SIGN II OR SIGN IV IS RIGHT	SIGN III IS WRONG

VII	VIII	IX
THE LADY IS NOT IN ROOM I	THIS ROOM CONTAINS A TIGER AND ROOM IX IS EMPTY	THIS ROOM CONTAINS A TIGER AND VI IS WRONG

The prisoner studied the situation for a long while.

"The problem is unsolvable!" he exclaimed angrily. "That's not fair!"

"I know," laughed the king.

"Very funny!" replied the prisoner. "Come on, now, at least give me a decent clue: is Room Eight empty or not?"

The king was decent enough to tell him whether Room VIII was empty or not, and the prisoner was then able to deduce where the lady was.

Which room contained the lady?

◆ SOLUTIONS ◆

1 ◆ We are given that one of the two signs is true and the other false. Could it be that the first is true and the second false? Certainly not, because if the first sign is true, then the second sign must also be true—that is, if there is a lady in Room I and a tiger in Room II, then it is certainly the case that one of the rooms contains a lady and the other a tiger. Since it is not the case that the first sign is true and the second one false, then it must be that the second sign is true and the first one false. Since the second sign is true, then there really is a lady in one room and a tiger in the other. Since the

first sign is false, then it must be that the tiger is in Room I and the lady in Room II. So the prisoner should choose Room II.

2 ◆ If Sign II is false, then Room I contains a lady; hence at least one room contains a lady, which makes Sign I true. Therefore, it is impossible that both signs are false. This means that both signs are true (since we are given that they are either both true or both false). Therefore, a tiger is in Room I and a lady in Room II, so again the prisoner should choose Room II.

3 ◆ The king must have been in a generous mood this time, because both rooms contain ladies! We prove this as follows:

Sign I says in effect that *at least* one of the following alternatives holds: there is a tiger in Room I; there is a lady in Room II. (The sign does not preclude the possibility that both alternatives hold.)

Now, if Sign II is false, then a tiger is in Room I, which makes the first sign true (because the first alternative is then true). But we are given that it is not the case that one of the signs is true and the other one false. Therefore, since Sign II is true, both signs must be true. Since Sign II is true, there is a lady in Room I. This means that the first alternative of Sign I is false, but since at least one of the alternatives is true, then it must be the second one. So there is a lady in Room II also.

4 ◆ Since the signs say the same thing, they are both true or both false. Suppose they are true; then both rooms contain ladies. This would mean in particular that Room II contains a lady. But we have been told that if Room II contains a lady, the sign is false. This is a contradiction, so the signs are not

true; they are both false. Therefore, Room I contains a tiger and Room II contains a lady.

5 ◆ If the first room contains a tiger, we get a contradiction. Because if it does contain a tiger, then the first sign is false, which would mean that neither room contains a lady; both rooms would contain tigers. But we have been told that a tiger in the second room indicates that the second sign is true, which would mean that the other room contains a lady, contrary to the assumption that the first room contains a tiger. So it is impossible for the first room to contain a tiger; it must contain a lady. Therefore, what the second sign says is true, and the second room contains a tiger. So the first room contains a lady and the second room contains a tiger.

6 ◆ The first sign says, in effect, that either both rooms contain ladies or both contain tigers—that is the only way it could make no difference which room is picked.

Suppose the first room contains a lady. Then the first sign is true, which means the second room also contains a lady. Suppose, on the other hand, the first room contains a tiger. Then the first sign is false, which means that the two occupants are not the same, so again the second occupant is a lady. This proves that Room II must contain a lady regardless of what is in Room I. Since Room II contains a lady, Sign II is false and Room I must contain a tiger.

7 ◆ The first sign says in effect that the two occupants are different (one a lady and the other a tiger), but doesn't say which room contains which. If Room I's occupant is a lady, the sign is true; hence Room II must contain a tiger. If, on the other hand, Room I's occupant is a tiger, then the first sign is false, which means that the two occupants are not really dif-

ferent, so Room II must also contain a tiger. Therefore, Room II definitely contains a tiger. This means that the second sign is true, so a lady must be in the first room.

8 ◆ Suppose the top sign, THIS ROOM CONTAINS A TIGER, were on the door of Room I. If a lady is in the room, the sign is false, violating the conditions given by the king. If a tiger is in the room, the sign is true, which again violates the king's conditions. So that sign can't be on the first door; it must be on the second. This means the other sign is to be put on the first door.

The sign belonging on the first door thus reads: BOTH ROOMS CONTAIN TIGERS. So the first room can't contain a lady, or the sign would be true, which would mean that both rooms contain tigers—an obvious contradiction. Therefore, the first room contains a tiger. From this it follows that the sign is false, so the second room must contain a lady.

9 ◆ Signs II and III contradict each other, so at least one of them is true. Since at most one of the three signs is true, then the first one must be false, so the lady is in Room I.

10 ◆ Since the sign of the room containing the lady is true, then the lady certainly can't be in Room II. If she is in Room III, then all three signs must be true, which is contrary to the given condition that at least one sign is false. Therefore, the lady is in Room I (and sign II is true and sign III is false).

11 ◆ Since the sign on the door of the room containing the lady is true, then the lady cannot be in Room III.

Suppose she is in Room II. Then sign II would be true; hence the tiger would be in Room I and Room III would be empty. This would mean that the sign on the door of the tiger's room would be true, which is not possible. Therefore, the lady is in Room I; Room III must then be empty, and the tiger is in Room II.

12 ◆ If the king had told the prisoner that Room VIII was empty, it would have been impossible for the prisoner to have found the lady. Since the prisoner did deduce where the lady was, the king must have told him that Room VIII was not empty, and the prisoner reasoned as follows:

"It is impossible for the lady to be in Room Eight, for if she were, Sign Eight would be true, but the sign says a tiger is in the room, which would be a contradiction. Therefore, Room Eight does not contain the lady. Also, Room Eight is not empty; therefore, Room Eight must contain a tiger. Since it contains a tiger, the sign is false. Now, if Room Nine is empty, then Sign Eight would be true; therefore, Room Nine cannot be empty.

"So, Room Nine is also not empty. It cannot contain the lady, or the sign would be true, which would mean that the room contains a tiger; this means Sign Nine is false. If Sign Six is really wrong, then Sign Nine would be true, which is not possible. Therefore, Sign Six is right.

"Since Sign Six is right, then Sign Three is wrong. The only way Sign Three can be wrong is that Sign Five is Wrong and Sign Seven is right. Since Sign Five is wrong, then Sign Two and Sign Four are both wrong. Since Sign Four is wrong, then Sign One must be right.

"Now I know which signs are right and which signs are wrong—namely:

"1 – Right	4 – Wrong	7 – Right
2 – Wrong	5 – Wrong	8 – Wrong
3 – Wrong	6 – Right	9 – Wrong

"I now know that the lady is in either Room One, Room Six, or Room Seven, since the others all have false signs. Since Sign One is right, the lady can't be in Room Six. Since Sign Seven is right, the lady can't be in Room One. Therefore, the lady is in Room Seven."

◆ 3 ◆

The Asylum of Doctor Tarr and Professor Fether

Inspector Craig of Scotland Yard was called over to France to investigate eleven insane asylums where it was suspected that something was wrong. In each of these asylums, the only inhabitants were patients and doctors—the doctors constituted the entire staff. Each inhabitant of each asylum, patient or doctor, was either sane or insane. Moreover, the sane ones were *totally* sane and a hundred percent accurate in all their beliefs; all true propositions they knew to be true and all false propositions they knew to be false. The insane ones were totally inaccurate in their beliefs; all true propositions they believed to be false and all false propositions they believed to be true. It is to be assumed also that all the inhabitants were always honest—whatever they said, they really believed.

1 ◆ The First Asylum

In the first asylum Craig visited, he spoke separately to two inhabitants whose last names were Jones and Smith.

"Tell me," Craig asked Jones, "what do you know about Mr. Smith?"

"You should call him *Doctor* Smith," replied Jones. "He is a doctor on our staff."

Sometime later, Craig met Smith and asked, "What do you know about Jones? Is he a patient or a doctor?"

"He is a patient," replied Smith.

The inspector mulled over the situation for a while and then realized that there was indeed something wrong with this asylum: either one of the doctors was insane, hence shouldn't be working there, or, worse still, one of the patients was sane and shouldn't be there at all.

How did Craig know this?

2 • The Second Asylum

In the next asylum Craig visited, one of the inhabitants made a statement from which the inspector could deduce that the speaker must be a sane patient, hence did not belong there. Craig then took steps to have him released.

Can you supply such a statement?

3 • The Third Asylum

In the next asylum, an inhabitant made a statement from which Craig could deduce that the speaker was an insane doctor. Can you supply such a statement?

4 • The Fourth Asylum

In the next asylum, Craig asked one of the inhabitants, "Are you a patient?" He replied, "Yes."

Is there anything necessarily wrong with this asylum?

5 • The Fifth Asylum

In the next asylum, Craig asked one of the inhabitants, "Are you a patient?" He replied, "I believe so."

Is there anything necessarily wrong with this asylum?

6 • The Sixth Asylum

In the next asylum Craig visited, he asked an inhabitant, "Do you believe you are a patient?" The inhabitant replied, "I believe I do."

Is there anything necessarily wrong with this asylum?

7 • The Seventh Asylum

Craig found the next asylum more interesting. He met two inhabitants, A and B, and found out that A believed that B was insane and B believed that A was a doctor. Craig then took measures to have one of the two removed. Which one, and why?

8 • The Eighth Asylum

The next asylum proved to be quite a puzzler, but Craig finally managed to get to the bottom of things. He found out that the following conditions prevailed:

1. Given any two inhabitants, A and B, either A trusts B or he doesn't.

2. Some of the inhabitants are *teachers* of other inhabitants. Each inhabitant has at least one teacher.

3. No inhabitant A is willing to be a teacher of an inhabitant B unless A believes that B trusts himself.

4. For any inhabitant A there is an inhabitant B who trusts all and only those inhabitants who have at least one teacher who is trusted by A. (In other words, for any inhabitant X, B trusts X if A trusts some teacher of X, and B doesn't trust X unless A trusts some teacher of X.)

5. There is one inhabitant who trusts all the patients but does not trust any of the doctors.

Inspector Craig thought this over for a long time and was finally able to prove that either one of the patients was sane or one of the doctors was insane. Can you find the proof?

9 • The Ninth Asylum

In this asylum, Craig interviewed four inhabitants: A, B, C, and D. A believed that B and C were alike as far as their sanity was concerned. B believed that A and D were alike as far as their sanity was concerned. Then Craig asked C, "Are you and D both doctors?" C replied, "No."

Is there anything wrong with this asylum?

10 • The Tenth Asylum

Inspector Craig found this case particularly interesting, though difficult to crack. The first thing he discovered was that the asylum's inhabitants had formed various committees. Doctors and patients, he learned, could serve on the same committee and sane and insane persons might be on the same committee. Then Craig found out the following facts:

1. All patients formed one committee.

2. All doctors formed one committee.

3. Each inhabitant had several friends in the asylum, and among them one who was his best friend. Also, each inhabitant had several enemies in the asylum, and among them one called his worst enemy.

4. Given any committee, C, all inhabitants whose best friend was on C formed a committee, and all inhabitants whose worst enemy was on C also formed a committee.

5. Given any two committees—Committee 1 and Committee 2—there was at least one inhabitant, D, whose best friend believed that D was on Committee 1 and whose worst enemy believed that D was on Committee 2.

Putting all these facts together, Craig found an ingenious proof that either one of the doctors was insane or one of the patients was sane. How did Craig know this?

11 • An Added Puzzle

Craig lingered for a while in this last asylum, because certain other questions caught his theoretical fancy. For example, he was curious to know whether all sane inhabitants formed a committee and all insane inhabitants formed a committee. He could not settle these questions on the basis of facts 1, 2, 3, 4, and 5, but he was able to prove—and just on the basis of 3, 4, and 5—that it was not possible for *both* of these groups to have formed committees. How did he prove this?

12 • Another Puzzle About the Same Asylum

Finally, Craig was able to prove something else about this same asylum. He regarded it as quite significant, and in fact it simplified the solutions of the last two problems. It is, namely, that given any two committees—Committees 1 and 2—there must be an inhabitant E and an inhabitant F who believe as follows: E that F serves on Committee 1, and F that E serves on Committee 2. How did Craig prove this?

13 • The Asylum of Doctor Tarr and Professor Fether

The last asylum Craig visited he found to be the most bizarre of all. This asylum was run by two doctors named Doctor Tarr and Professor Fether. There were other doctors on the staff as well. Now, an inhabitant was called *peculiar* if he believed that he was a patient. An inhabitant was called *special* if all patients believed he was peculiar and no doctor believed he was peculiar. Inspector Craig found out that at least one inhabitant was sane and that the following condition held:

Condition C: Each inhabitant had a best friend in the asylum. Moreover, given any two inhabitants, A and B, if A believed that B was special, then A's best friend believed that B was a patient.

Shortly after this discovery, Inspector Craig had private interviews with Doctor Tarr and Professor Fether. Here is the interview with Doctor Tarr:

Craig: Tell me, Doctor Tarr, are all the doctors in this asylum sane?

Tarr: Of course they are!

Craig: What about the patients? Are they all insane?

Tarr: At least one of them is.

The second answer struck Craig as a surprisingly modest claim! Of course, if all the patients are insane, then it certainly is true that at least one is. But why was Doctor Tarr being so cautious? Craig then had his interview with Professor Fether, which went as follows:

Craig: Doctor Tarr said that at least one patient here is insane. Surely that is true, isn't it?

Professor Fether: Of course it is true! All the patients in this asylum are insane! What kind of asylum do you think we are running?

Craig: What about the doctors? Are they all sane?

Professor Fether: At least one of them is.

Craig: What about Doctor Tarr? Is he sane?

Professor Fether: Of course he is! How dare you ask me such a question?

At this point, Craig realized the full horror of the situation! What was it?

(Those who have read "The System of Doctor Tarr and Professor Fether," by Edgar Allan Poe, will probably guess the solution before they prove it is correct. See remarks following the solution.)

• SOLUTIONS •

1 • We will prove that either Jones or Smith (we do not know which) must be either an insane doctor or a sane patient (but again we don't know which).

Jones is either sane or insane. Suppose he is sane. Then his belief is correct; hence Smith really is a doctor. If Smith is insane, then he is an insane doctor. If Smith is sane, then his belief is correct, which means that Jones is a patient and hence a sane patient (since we are assuming Jones to be sane). This proves that if Jones is sane, then either he is a sane patient or Smith is an insane doctor.

Suppose Jones is insane. Then his belief is wrong, which makes Smith a patient. If Smith is sane, then he is a sane patient. If Smith is insane, then his belief is wrong, which makes Jones a doctor, hence an insane doctor. This proves that if Jones is insane, then either he is an insane doctor or Smith is a sane patient.

To summarize, if Jones is sane, then either he is a sane patient or Smith is an insane doctor. If Jones is insane, then either he is an insane doctor or Smith is a sane patient.

2 ◆ Many solutions are possible. The simplest I can think of is that the inhabitant said, "I am not a sane doctor." We then prove that the speaker must be a sane patient as follows:

An insane doctor could not hold the true belief that he is not a sane doctor. A sane doctor could not hold the false belief that he is not a sane doctor. An insane patient could not hold the true belief that he is not a sane doctor (an insane patient is in fact *not* a sane doctor). So the speaker was a sane patient, and his belief that he was not a sane doctor was correct.

3 ◆ One statement which would work is: "I am an insane patient." A sane patient could not hold the false belief that he is an insane patient. An insane patient could not hold the true belief that he is an insane patient. Therefore, the speaker was not a patient; he was a doctor. A sane doctor could never believe that he is an insane patient. So the speaker was an insane doctor, who held the false belief that he was an insane patient.

4 ◆ The speaker believes that he is a patient. If he is sane, then he really is a patient; hence he is a sane patient and

shouldn't be in the asylum. If he is insane, his belief is wrong, which means that he is not a patient but a doctor; hence he is an insane doctor and shouldn't be on the staff. It is not possible to tell whether he is a sane patient or an insane doctor, but in neither case should he be in this asylum.

5 ◆ This is a very different situation! Just because the speaker *says* that he believes he is a patient does not necessarily mean that he *does* believe he is a patient! Since he says he believes he is a patient, and he is honest, then he ***believes*** that he believes he is a patient. Suppose he is insane. Then his beliefs are all wrong—even those about his own beliefs—so his believing that he believes he is a patient indicates that it is false that he believes he is a patient, and therefore he actually believes that he is a doctor. Since he is insane and believes he is a doctor, then he is in fact a patient. So if he is insane, he is an insane patient. On the other hand, suppose he is sane. Since he believes that he believes he is a patient, then it is true that he believes he is a patient. Since he believes he is a patient, then he is a patient. So, if he is sane, then again he is a patient. We see, therefore, that he could be either a sane patient or an insane patient, and we have no grounds for finding anything wrong with this asylum.

More generally, let us note the following basic facts: If an inhabitant of this asylum believes something, then that something is true or false depending on whether the believer is sane or insane. But if an inhabitant believes that he believes something, then the something must be true, regardless of whether the believer is sane or insane. (If he is insane, the two beliefs cancel each other, analogously to the negative of a negative making a positive.)

6 ◆ In this case, the speaker didn't claim that he was a patient nor that he believed he was a patient; he claimed that

he believed that he believed he was a patient. Since he believed what he claimed, then he believed that he believed that he believed that he was a patient. The first two beliefs cancel each other out (see last paragraph of the solution to the last problem), so in fact he believed that he was a patient. The problem then reduces to that of the fourth asylum, which we have already solved (the speaker must be either a sane patient or an insane doctor).

7 • Craig had A removed. *Reason:* Suppose A is sane. Then his belief that B is insane is correct. Since B is insane, his belief that A is a doctor is wrong, so A is a sane patient and should be removed. Suppose, on the other hand, that A is insane. Then his belief that B is insane is wrong, so B is sane. Then B's belief that A is a doctor is correct, so in this case A is an insane doctor who should be removed.

Nothing at all can be deduced about B.

8 • By condition 5, there is an inhabitant—call him Arthur—who trusts all patients but no doctors. By condition 4, there is an inhabitant—call him Bill—who trusts just those inhabitants who have at least one teacher who is trusted by Arthur. This means that for any inhabitant X, if Bill trusts X, then Arthur trusts at least one teacher of X, and if Bill doesn't trust X, then Arthur trusts no teacher of X. Since being trusted by Arthur is the same thing as being a patient (by condition 5), then we can rephrase the last sentence as follows: For any inhabitant X, if Bill trusts X, then at least one teacher of X is a patient, and if Bill doesn't trust X, then no teacher of X is a patient. Now, since this holds for *every* inhabitant X, then it also holds when X is Bill himself. Therefore, we know the following:

(1) If Bill trusts himself, then Bill has at least one teacher who is a patient.

(2) If Bill doesn't trust himself, then no teacher of Bill is a patient.

There are two possibilities: either Bill trusts himself or he doesn't. Let us now see what is implied by each case.

Case 1—Bill trusts himself: Then Bill has at least one teacher—call him Peter—who is a patient. Since Peter is a teacher of Bill, then Peter believes that Bill trusts himself (this according to condition 3). Well, Bill does trust himself, so Peter believes truly and is sane. Therefore, Peter is a sane patient and should not be in this asylum.

Case 2—Bill doesn't trust himself: In this case, none of Bill's teachers is a patient. Yet Bill, like every other inhabitant, has at least one teacher—call him Richard. Then Richard must be a doctor. Since Richard is a teacher of Bill, then Richard believes that Bill trusts himself. His belief is wrong; therefore Richard is insane. So Richard is an insane doctor and shouldn't be on the staff.

To summarize: if Bill trusts himself, then at least one patient is sane. If Bill doesn't trust himself, then at least one doctor is insane. Since we don't know whether Bill trusts himself or not, we don't know just what is wrong with this asylum—whether there is a sane patient or an insane doctor.

9 ◆ We shall first prove that C and D are necessarily alike as far as their sanity is concerned.

Suppose A and B are both sane. Then B and C are really alike, and A and D are really alike. This implies that all four are sane; hence in this case C and D are both sane, and thus alike. Now suppose A and B are both insane. Then B and C are different, and so are A and D; hence C and D are both sane, and so again alike. Now suppose A is sane and B is insane. Then B and C are alike, so C is insane, but A and D are different, which means D is also insane. Lastly, suppose A is insane and B is sane. Then B and C are different and so C is insane, but A and D are alike; hence D is also insane.

In summary, if A and B are alike, then C and D are both sane; if A and B are different, C and D are both insane.

Thus, we have established that C and D are either both sane or both insane. Suppose they are both sane. Then C's statement that he and D are not both doctors was true, which means that at least one is a patient, hence a sane patient. If C and D are both insane, then C's statement was false, which means they are both doctors, hence both insane doctors. Therefore, this asylum contains at least one sane patient, or else at least two insane doctors.

10, 11, 12 ◆ First read problems 11 and 12, because the easiest way to solve Problem 10 is to start with Problem 12.

Before we begin, let me point out a useful principle: Suppose we have two statements, X and Y, which are known to be either both true or both false. Then any inhabitant of the asylum, if he believes one of the statements, must also believe the other. *Reason:* If the statements are both true, then any inhabitant who believes one of them must be sane, hence must also believe the other, which is also true. If the statements are both false, then any inhabitant who believes one of them must be insane, and must also believe the other, since it is also false.

Now let us solve Problem 12: Take any two committees, Committee 1 and Committee 2. Let U be the group of all inhabitants whose worst enemy belongs to Committee 1, and let V be the group of all inhabitants whose best friend belongs to Committee 2. According to Fact 4, both U and V are committees. Therefore, according to Fact 5, there is some inhabitant—call him Dan—whose best friend—call him Edward—believes Dan is on U, and whose worst enemy—call him Fred—believes Dan is on V. Thus, Edward believes Dan is on Committee U and Fred believes Dan is on Committee V. Now, by the definition of U, to say that Dan is on U is to say that his worst enemy, Fred, is on Committee 1; in other words, the two statements "Dan is on U" and "Fred is on Committee 1" are either both true or both false. Since Edward believes the one—namely, that Dan is on U—then he must also believe the other—namely, that Fred is on Committee 1 (recall our preliminary principle!). So Edward believes Fred is on Committee 1.

Fred, on the other hand, believes that Dan is on Committee V. Now, Dan is on V only if his friend Edward is on Committee 2 (by the defintion of V); in other words, these two facts are either both true or both false. Then, since Fred believes Dan is on V, Fred must also believe that Edward is on Committee 2.

Thus we have two inhabitants, Edward and Fred, with these beliefs: Edward that Fred is on Committee 1, and Fred that Edward is on Committee 2. This solves Problem 12.

To solve Problem 10, let us now take as Committee 1 the group of all patients and as Committee 2 the group of all doctors, which are committees according to Facts 1 and 2. According to the solution of Problem 12, there exist inhabitants—Edward and Fred—who believe the following: Edward, that Fred is on Committee 1 of all patients; and Fred, that Edward is on Committee 2 of all doctors. In other words, Edward believes that Fred is a patient and Fred believes that Edward is a doctor. Then, according to Problem 1 (using the names *Edward* and *Fred* rather than *Jones* and *Smith*), one of the two, Edward or Fred (we don't know which), must be either an insane doctor or a sane patient. So something is definitely wrong with this asylum.

As for Problem 11, suppose that the group of all sane inhabitants and the group of all insane inhabitants *are* both committees, Committees 1 and 2 respectively. Then, according to Problem 12, inhabitants Edward and Fred would believe: (a) Edward, that Fred is sane—in other words, a member of Committee 1; (b) Fred, that Edward is insane—and thus a member of Committee 2. This is impossible, because if Edward is sane, his belief is correct, which means Fred is sane; hence Fred's belief is correct, which means Edward is insane. So if Edward is sane, he is also insane, which is impossible. On the other hand, if Edward is insane, his belief about Fred is wrong, which means Fred is insane; hence Fred's belief about Edward is wrong, which means Edward is sane. So if Edward is insane, he is also sane, which again is impossible. Therefore, the assumption that the group of sane inhabitants and the group of insane inhabitants are *both* committees leads to a contradiction. Therefore, it cannot be that both these groups are committees.

13 • What Craig realized, to his horror, was that in this asylum, all the doctors were insane and all the patients were sane! He reasoned this out in the following manner:

Even before his interviews with Doctor Tarr and Professor Fether, he knew that there was at least one sane inhabitant, A. Now let B be A's best friend. By Condition C, if A believes that B is special, then A's best friend believes that B is a patient. Since A's best friend is B, then if A believes that B is special, B believes that B is a patient. In other words, if A believes that B is special, then B is peculiar. Since A is sane, then A's believing that B is special is tantamount to B's actually being special. Therefore, we have the following *key fact:*

If B is special, then B is peculiar.

Now, either B is peculiar or he isn't. If he is peculiar, then he believes he is a patient, and therefore (see Problem 4) he must be either an insane doctor or a sane patient; either way, B shouldn't be in the asylum. But suppose B is not peculiar, what then? Well, if B is not peculiar, he is not special either, because, in accordance with the "key fact," B can only be special if he is also peculiar. So B is neither peculiar nor special. Since he is not special, then the two assumptions that all patients believe he is peculiar and that no doctor believes he is peculiar cannot both be true, which means that at least one of them is false. Suppose the first assumption is false. Then at least one patient, P, does not believe that B is peculiar. If P were insane, then he would believe that B is peculiar (since B isn't); therefore, P is sane. This means that P is a sane patient. If the second assumption is false, then at least one doctor, D, believes that B is peculiar. Then D must be insane (since B is not peculiar); so D is an insane doctor.

To summarize: If B is peculiar, then he is either a sane patient or an insane doctor. If B is not peculiar, then either some sane patient, P, doesn't believe that B is peculiar, or some insane doctor, D, does believe B is. Therefore, this asylum must contain either a sane patient or an insane doctor.

As I said, Craig realized all this before his interviews with Doctor Tarr and Professor Fether. Now, Doctor Tarr believes that all the doctors are sane, and Professor Fether believes that all the patients are insane. They cannot both be right (as we have proved); hence at least one of them is insane. Also, Professor Fether believes that Doctor Tarr is sane. If Professor Fether is sane, he has to be right, and Doctor Tarr would also be sane, which we know is not true. Therefore, Professor Fether must be insane. Then his belief that Doctor Tarr is sane is wrong, so Doctor Tarr is also insane. This proves that Doctor Tarr and Professor Fether are both insane.

Since Doctor Tarr is insane and believes that at least one

patient is insane, then in fact all the patients must be sane. Since Professor Fether is insane and believes that at least one doctor is sane, then in fact all the doctors are insane. This proves that all the patients are sane and all the doctors are insane.

Remarks: This puzzle, of course, was suggested by Edgar Allan Poe's story "The System of Doctor Tarr and Professor Fether," in which the patients of a lunatic asylum managed to overcome all the doctors and staff, put them, tarred and feathered, in the patients' cells, and assumed their rôles.

◆ 4 ◆

Inspector Craig Visits Transylvania

A week after these last adventures, Craig was preparing to return to London when he suddenly received a wire from the Transylvanian government, urgently requesting him to come to Transylvania to help solve some baffling cases of vampirism. Now, as I explained in my previous book of logic puzzles, *What Is the Name of This Book?*, Transylvania is inhabited by both vampires and humans; the vampires always lie and the humans always tell the truth. However, half the inhabitants, both human and vampire, are insane and totally deluded in their beliefs (just like the mad inhabitants of the asylum of Doctor Tarr and Professor Fether)—all true propositions they believe false and all false propositions they believe true. The other half of the inhabitants are completely sane and totally accurate in their judgments (just like the sane inhabitants of the asylums in Chapter 3)—all true statements they know to be true and all false statements they know to be false.

Of course, the logic of Transylvania is much more complicated than that of the lunatic asylums, because in the latter, the inhabitants are at least honest and make false statements only out of delusion, never out of malice. But when a Transylvanian makes a false statement, it could be either out of delusion or out of malice. Sane humans and insane vampires both make only true statements; insane humans and sane vampires make only false statements. For example, if you ask a Transylvanian whether the earth is round (as opposed to flat), a sane human knows the earth is round and will truthfully say so. An insane human believes the earth is not round, and will then truthfully express his belief and say it is not round. A sane vampire knows the earth is round, but will then lie and say it isn't. But an insane vampire believes the earth is not round and then lies and says it is round. Thus an insane vampire responds the same way to any question as a sane human, and an insane human the same way as a sane vampire.

It was fortunate that Craig was as well versed in vampirism as in logic (the general range of Craig's interests and knowledge was quite remarkable altogether). When he arrived in Transylvania, he was informed by the authorities (all of whom were sane humans) that there were ten cases with which they needed help, and he was requested to take charge of the investigations.

THE FIRST FIVE CASES

Each of these cases involved two inhabitants. In each case, it was already known that one of them was a vampire and the other was human, but it was not known which was which (or perhaps I should say, which was witch). Nothing was known, except in Case 5, about the sanity of either.

1 ♦ The Case of Lucy and Minna

The first case involved two sisters named Lucy and Minna, and Craig had to determine which one of them was a vampire. As indicated above, nothing was known about the sanity of either. Here is the transcript of the investigation:

Craig (to Lucy): Tell me something about yourselves.

Lucy: We are both insane.

Craig (to Minna): Is that true?

Minna: Of course not!

From this, Craig was able to prove to everyone's satisfaction which of the sisters was the vampire. Which one was it?

2 ♦ Case of the Lugosi Brothers

The next case was that of the Lugosi brothers. Both had the first name of Bela. Again, one was a vampire and one was not. They made the following statements:

Bela the Elder: I am human.

Bela the Younger: I am human.

Bela the Elder: My brother is sane.

Which one is the vampire?

3 ♦ The Case of Michael and Peter Karloff

The next case involved another pair of brothers, Michael and Peter Karloff. Here is what they said:

Michael Karloff: I am a vampire.

Peter Karloff: I am human.

Michael Karloff: My brother and I are alike as far as our sanity goes.

Which one is the vampire?

4 • The Case of the Turgeniefs

The next case involved a father and son whose surname was Turgenief. Here is the transcript of the interrogation:

Craig (to the father): Are you both sane or both insane, or are you different in this respect?

Father: At least one of us is insane.

Son: That is quite true!

Father: I, of course, am not a vampire.

Which one is the vampire?

5 • The Case of Karl and Martha Dracula

The last case of this group involved a pair of twins, Karl and Martha Dracula (no relation to the count, I can assure you!). The interesting thing about this case is that not only was it already known that one of them was human and the other a vampire, but it was also known that one of the two was sane and the other insane, although Craig had no idea which was which. Here is what they said:

Karl: My sister is a vampire.

Martha: My brother is insane!

Which one is the vampire?

FIVE MARRIED COUPLES

The next five cases each involved a married couple. Now (as you may or may not know), in Transylvania it is illegal for humans and vampires to intermarry, hence any married couple there are either both humans or both vampires. In these cases, as in Problems 1 through 4, nothing was known about the sanity of either person.

6 • The Case of Sylvan and Sylvia Nitrate

The first case in this group was that of Sylvan and Sylvia Nitrate. As already explained, they are either both humans or both vampires. Here is the transcript of Craig's interrogation:

Craig (to Mrs. Nitrate): Tell me something about yourselves.

Sylvia: My husband is human.

Sylvan: My wife is a vampire.

Sylvia: One of us is sane and one of us is not.

Are they humans or vampires?

7 • The Case of George and Gloria Globule

The next case involved the Globules.

Craig: Tell me something about yourselves.

Gloria: Whatever my husband says is true.

George: My wife is insane.

Craig did not feel that the husband's remark was overly gallant; nevertheless, these two testimonies were sufficient to solve the case.

Is this a human or a vampire couple?

8 • The Case of Boris and Dorothy Vampyre

"It is important," said the Transylvanian chief of police to Inspector Craig, "not to let the last name of the suspects prejudice the issue."

Here are the answers they gave:
Boris Vampyre: We are both vampires.
Dorothy Vampyre: Yes, we are.
Boris Vampyre: We are alike, as far as our sanity goes.
What kind of couple are we dealing with?

9 ◆ The Case of Arthur and Lillian Sweet

The next case involved a foreign couple (foreign to Transylvania, that is) named Arthur and Lillian Sweet. Here is their testimony:
Arthur: We are both insane.
Lillian: That is true.
What are Arthur and Lillian?

10 ◆ The Case of Luigi and Manuella Byrdcliffe

Here is the testimony of the Byrdcliffes:
Luigi: At least one of us is insane.
Manuella: That is not true!
Luigi: We are both human.
What are Luigi and Manuella?

TWO UNEXPECTED PUZZLES

11 ◆ The Case of A and B

Inspector Craig was relieved that all these unpleasant cases were over and was packing his things for his return to London, when quite unexpectedly a Transylvanian official burst into his room, begging him to stay just one more day to help solve a new case that had just come up. Well, Craig certainly did not relish the idea; still, he felt it his duty to assist where possible, and he consented.

It appeared that two suspicious-looking characters had just been picked up by the Transylvanian police. They both happened to be prominent persons, and Craig has requested that their names and sexes be withheld, so I shall just call them A and B. In contrast to the previous ten trials, nothing was known in advance concerning any relationship between them; they might both be vampires or both be human, or one could be a vampire and the other human. Also, they could be both sane or both insane, or one could be sane and the other insane.

At the trial, A stated that B was sane, and B claimed that A was insane. Then A claimed that B was a vampire, and B declared that A was human.

What can be deduced about A and B?

12 • Two Transylvanian Philosophers

Happy that these weird trials were over at last, Craig was comfortably seated in a Transylvanian railroad station awaiting the train that would take him out of the country. He so looked forward to being back in London! Just then he overheard a dispute between two Transylvanian philosophers, who were eagerly discussing the following problem:

Suppose there is a pair of identical Transylvanian twins, one of whom is known to be a sane human and the other an insane vampire. And suppose you meet one of them alone and wish to find out which one he is. Can any amount of yes/no questions suffice to do this? The first philosopher maintained that no number of questions could possibly accomplish this, since either one would give the same answer as the other to any question asked. That is, given any question, if its correct answer is yes, the sane human will know the answer is yes and will truthfully answer yes; whereas the insane vampire will believe the answer is no and then lie and say yes. Similarly, if the correct answer to the question is no, then the sane human will answer no and the insane vampire, thinking the answer is yes, will lie and also say no. Therefore, the two brothers are indistinguishable in their outward verbal behavior, even though their minds work entirely differently. So, the first philosopher argued, no questions could tell them apart—unless, perhaps, given with a lie detector.

The second philosopher disagreed. Actually, he did not present any arguments to support his position; all he said was, "Let me interrogate one of those two brothers, and I'll tell you which one he is!"

Craig was curious to hear the outcome of the dispute, but just then his train pulled in and the philosophers did not board it.

Inspector Craig sat in his carriage for some time pondering as to which philosopher was right. He eventually realized that it was the second philosopher: if you met one of the twins, you could indeed find out by yes/no questions which of them you were addressing, and no lie detector was necessary. This then leaves two problems:

(1) What is the smallest number of questions you need to ask?

(2) More interesting yet, just what is wrong with the first philosopher's argument?

• SOLUTIONS •

There is a principle that will apply in several of the solutions that follow and which we will establish in advance—namely, that if a Transylvanian says he is human, then he must be sane, and if he says he is a vampire, then he must be insane. The reason is this: Suppose he says he is human. Now, his statement is either true or false. If his statement is true, then he really is human, but the only humans who make true statements are sane humans so in this case he is sane. If, on the other hand, his statement is false, then he is really a vampire, but the only vampires who make false statements are sane vampires (insane vampires make true statements, just like sane humans), so again he is sane. This proves that when a Transylvanian claims to be human, he must be sane, regardless of whether he is really human or not.

Suppose a Transylvanian claims to be a vampire; what follows? Well, if his claim is true, then he really is a vampire, but the only vampires who make true claims are insane vampires. If his claim is false, then he is in fact human, but the only humans who make false claims are insane humans; so in this case he is also insane. Thus, any Transylvanian who claims to be a vampire is insane.

We trust that the reader can verify for himself the fact that any Transylvanian who claims to be sane must in fact be human, and any Transylvanian who claims to be insane must in fact be a vampire.

Now let us turn to the solutions of the problems.

1 • Lucy's statement is either true or false. If it is true, then both sisters are really insane; hence Lucy is insane, and the

only insane Transylvanian who can make a true statement is an insane vampire. So, if Lucy's statement is true, then Lucy is a vampire.

Suppose Lucy's statement is false. Then at least one of the sisters is sane. If Lucy is sane, then, since she has made a false statement, she must be a vampire (because sane humans make only true statements). Suppose Lucy is insane. Then it must be Minna who is sane. Also, Minna, by contradicting Lucy's false statement, has made a true statement. Therefore, Minna is sane and has made a true statement; so Minna is human, and again Lucy must be the vampire.

This proves that regardless of whether Lucy's statement is true or false, Lucy is the vampire.

2 ◆ We have already established the principle that any Transylvanian who says he is human must be sane and any Transylvanian who says he is a vampire must be insane (see discussion prefacing the solutions). Now, both the Lugosi brothers claim to be human; therefore, they are both sane. Therefore, Bela the Elder makes a true statement when he says that his brother is sane. So Bela the Elder is both sane and makes true statements; hence he is human. Therefore, it is Bela the Younger who is the vampire.

3 ◆ Since Michael claims to be a vampire, he is insane, and since Peter claims to be human, he is sane. So Michael is insane and Peter is sane; thus the two brothers are *not* alike as far as their sanity goes. Therefore, Michael's second statement is false, and since Michael is insane, he must be human (insane vampires don't make false statements!). Therefore, Peter is the vampire.

4 ◆ Father and son agree in answering the question about their sanity. This means that they either both make true

statements or both make false statements. But, since only one of them is human and the other is a vampire, they must necessarily be different as regards their sanity: If they are both sane, the one who is human would make true statements and the vampire would make false statements, and they could never agree; if they are both insane, the human would make false statements and the vampire would make true statements, and again they could not agree. Therefore, it is really true that at least one of them is insane. This proves that both of them make true statements. Then, since the father says he is not a vampire, he really isn't. So it is the son who is the vampire.

5 ◆ Suppose Martha is the vampire. Then Karl is human, and also Karl has made a true statement; hence Karl in this case has to be a sane human. This would make Martha an insane vampire, since, as we have been told, Karl and Martha are different as regards their sanity. But then Martha, an insane vampire, would have made a false statement—that Karl is insane—which insane vampires cannot do. Therefore, the assumption that Martha is a vampire leads to a contradiction. So it is Karl who is the vampire.

We can also determine their sanity or lack of it: Karl has made a false statement; hence, being a vampire, he is sane. But then Martha has also made a false statement; hence, being human, she is insane. So the complete answer is that Karl is a sane vampire and Martha is an insane human; Karl is lying when he says that his sister is a vampire, and Martha is deluded when she says that her brother is insane. (Quite a pair, even for Transylvania!)

6 ◆ Now we are in the situation where either both are vampires or both are human. Therefore, the first two statements cannot both be right, nor can they both be wrong (for if they

are both wrong, Sylvan would be a vampire and Sylvia would be human). So one of the two statements is right and one is wrong. This means that one of the two people is sane and the other insane (because if they were both sane, their statements would both be right if they were human, and both wrong if they were vampires). Therefore, Sylvia is right when she says that one of the two is sane and the other insane. This means that Sylvia makes true statements. Therefore her statement that her husband is human is true. This means that they are both human (and, incidentally, Sylvia is sane and Sylvan insane).

7 ◆ Gloria, in saying that whatever her husband says is true, is assenting to his claim that she is insane; in other words, Gloria is indirectly claiming to be insane. Only vampires can make such a claim (as we proved in the discussion preceding the solutions); hence Gloria must be a vampire. Therefore, they are both vampires.

8 ◆ Suppose they are human. Then their statements that they are both vampires are false, which means they are insane humans. That would mean that they *are* alike as far as their sanity goes; hence Boris's second statement is true, which is not possible for an insane human. Therefore, they cannot be human; they are vampires (and insane ones).

9 ◆ Suppose they are human. A sane human couldn't possibly say that he/she and someone else are both insane hence they would both have to be insane humans. Then you would have insane humans making the true statement that they are both insane, which is not possible. Therefore, they cannot be human; they are vampires. (They could be either sane vampires who lie when they say that they are insane, or insane

vampires making the true statement that they are insane. Remember that insane vampires always make true statements, although they don't intend to!).

10 ◆ Luigi and Manuella contradict each other; one of them must be right and the other wrong. Therefore, one of them makes true statements and the other makes false statements. Since they are either both human or both vampires, then it must be true that at least one of them is insane, because if they are both sane, then they would either both make true statements if they were human or both make false statements if they were vampires. So Luigi is right when he says that at least one of the two is insane. Therefore, Luigi makes true statements, and when he says that they are both human, he is right about that, too. This proves that they are both human (and, incidentally, that Luigi is sane and Manuella insane).

11 ◆ Let us call a Transylvanian *reliable* if he makes correct statements and *unreliable* if he makes incorrect ones. Reliable Transylvanians are either sane humans or insane vampires; unreliable Transylvanians are either insane humans or sane vampires. Now, A claims that B is sane, and also that B is a vampire. A's two claims are either both true or both false. If they are both true, then B is a sane vampire, which means that B is unreliable. On the other hand, if A's claims are both false, then B must be an insane human, which again means that B is unreliable. So in either case (whether A's claims are both true or both false), B is unreliable. Hence B's claims are both false and A is neither insane nor human; therefore, A must be a sane vampire. This also means that A is unreliable; so A's claims are both false, which means that B must be an insane human. So the answer is that A is a sane vampire and B is an insane human.

Incidentally, this problem is only one out of sixteen of a similar nature that could be devised and that all have unique solutions. The *combination* of whatever two statements A makes about B (one about his sanity and one about his human or vampire nature) with any two statements of B about A (one about A's sanity and the other about his nature)—and there are sixteen possibilities for these four statements—will uniquely determine the precise character of both A and B. For example, if A says B is human and B is sane, and B says A is a vampire and A is insane, the solution will be that B is a sane human and A is an insane vampire. Again, suppose A says B is sane and B is a vampire, and B says A is insane and A is a vampire. What are A and B? *Answer:* A is a sane human and B is a sane vampire.

Have you seen how to solve each of these sixteen possible problems and why each one must have a unique solution? If not, look at it this way: A can make four possible pairs of statements about B—namely, (1) B is sane; B is human. (2) B is sane; B is a vampire. (3) B is insane; B is human. (4) B is insane; B is a vampire. In each of the four cases, we can determine whether or not B is reliable. In Case 1, B must be reliable regardless of whether A's statements are both true or both false—because if both are true, B is a sane human and hence reliable; if both are false, B is an insane vampire, hence again reliable. Likewise in Case 4, B must be reliable. In Cases 2 and 3, on the other hand, B must necessarily be unreliable. So from A's statements we can always determine the reliability of B. In a similar manner, from B's two statements we can determine the reliability of A. Then, when we know the respective reliabilities of A and B, we know which of all four statements are true and which are false, and the problem is then solved.

I might also remark that if, instead of A and B each making

two statements about the other, each made a *conjunction* of them, the problem would be unsolvable. If, for example, instead of making two separate statements—"B is sane," "B is a vampire"—A said, "B is a sane vampire," we could deduce nothing about the reliability of B; this is because if A's statement is correct, B *is* a sane vampire, but if A's statement is incorrect, B could be either an insane vampire or a sane human or an insane human.

12 ◆ One question is enough! All you need ask him is, "Are you human?" ("Are you sane?" would also work, and "Are you a sane human?" as well.) So suppose you ask him, "Are you human?" Well, if the one you are addressing is the sane human, he of course will answer yes. But suppose you are addressing the insane vampire. Being insane, he will erroneously believe he is human and then, being a vampire, will lie and say no. So the sane human will answer yes and the insane vampire will answer no. Therefore, if you get yes for an answer, you will know that he is the sane human, and if you get no for an answer, you will know that he is the insane vampire.

Now, more interesting yet, what was wrong with the first philosopher's argument? The first philosopher was certainly right in that if you ask the two brothers the same question, you will get the same answer. What the philosopher didn't realize was that if you ask, "Are *you* human?" to each of the two brothers, you are not actually asking the same question but rather two *different* questions, because the question contains the variable word *you,* whose meaning depends on the person to whom the question is addressed! So, even though you utter the same words when you put the question to two different people, you are really asking a different question in each case.

To look at it another way: Suppose the names of the two

brothers are known—John, say, is the name of the sane human, and Jim the name of the insane vampire. If I ask either brother, "Is John human?" both brothers will reply yes because I am now putting the *same* question to each; similarly, if I ask, "Is Jim human?" both brothers will answer no. But if I ask each brother, "Are *you* human?" I am really asking a different question in each case.

PART TWO

PUZZLES AND METAPUZZLES

◆ 5 ◆

The Island of Questioners

Somewhere in the vast reaches of the ocean, there is a very strange island known as the Island of Questioners. It derives its name from the fact that its inhabitants never make statements; they only ask questions. Then how do they manage to communicate? More on that later.

The inhabitants ask only questions answerable by yes or no. Each inhabitant is one of two types, A and B. Those of type A ask only questions whose correct answer is yes; those of type B ask only questions whose correct answer is no. For example, an inhabitant of type A could ask, "Does two plus two equal four?" But he could not ask whether two plus two equals five. An inhabitant of type B could not ask whether two plus two equals four, but he could ask whether two plus two equals five, or whether two plus two equals six.

1

Suppose you meet a native of this island, and he asks you, "Am I of type B?" What would you conclude?

2

Suppose, instead, he had asked you whether he is of type A. What would you have concluded?

3

I once visited this island and met a couple named Ethan and Violet Russell. I heard Ethan ask someone, "Are Violet and I both of type B?"

What type is Violet?

4

Another time I met two brothers whose first names were Arthur and Robert. Arthur once asked Robert, "Is at least one of us of type B?"

What types are Arthur and Robert?

5

Next I met a couple whose last name was Gordon. Mr. Gordon asked his wife, "Darling, are we of different types?"

What can be deduced about each?

6

Then I met a native whose last name was Zorn. He asked me, "Am I the type who could ask whether I am of type B?"

Can anything be deduced about Zorn, or is this story impossible?

7

Going from the sublime to the ridiculous, I came across a native who asked, "Am I the type who could ask the question I am now asking?"

Can anything be deduced about him?

8

I next came across a couple whose last name was Klink. Mrs. Klink asked her husband, "Are you the type who could ask me whether I am of type A?"

What can be deduced about Mr. and Mrs. Klink?

9

Then I met a couple named John and Betty Black. Betty asked John, "Are you the type who could ask whether at least one of us is of type B?"

What are John and Betty?

Remarks: The last two puzzles remind me of the title of a song I heard many years ago. It was part of a collection of songs all of which were sort of "spoofs" on psychoanalysis. This particular one was titled: "I can't get adjusted to the you who's gotten adjusted to me."

10

The next incident was really a logical tangle! I met three sisters named Alice, Betty, and Cynthia. Alice asked Betty, "Are you the type who could ask Cynthia whether she is the type who could ask you whether you two are of different types?"

As I walked away, I tried to puzzle this out and finally realized that it was possible to deduce the type of only one of the three girls. Which one, and which type is she?

A STRANGE ENCOUNTER

The next three exchanges I witnessed on the Island of Questioners were the most bizarre of all! Three patients from one of the insane asylums of Chapter 3 escaped and decided to pay a visit to the island. We recall that a patient from one of these asylums could be sane or insane and that the sane ones are totally accurate in all their beliefs, and the insane ones totally inaccurate in all their beliefs. We also recall that the patients, whether sane or insane, are always truthful; they never make statements unless they believe them to be true.

11

On the day after their arrival, one of the patients, whose name was Arnold, met a native of the island. The native asked him, "Do you believe I am of type B?"

What can be deduced about the native, and what can be deduced about Arnold?

12

The next day, another of the three patients, Thomas, had a long conversation with one of the natives (if you can call it a conversation—Thomas kept making statements and the native kept asking questions!). At one point the native asked Thomas, "Do you believe I am the type who could ask you whether you are insane?"

What can be deduced about the native, and what can be deduced about Thomas?

13

Several days later, I had a conversation with the third patient, whose name was William. William told me that on the preceding day he had overheard a conversation between Thomas and a native named Hal, in which Thomas said to Hal, "You are the type who could ask whether I believe you are of type B."

Can anything be deduced about either Thomas, Hal, or William?

WHO IS THE SORCERER?

At this point in my adventures, I still did not know whether Thomas was sane or insane, nor did I have much time to find out. The next day all three patients left the island. The last I heard, they had voluntarily returned to the asylum from which they escaped. They were evidently happy there, since they agreed unanimously that life outside the asylum was even crazier than life inside.

Well, it was a relief to have things back to normal on the Island of Questioners. Then I heard a rumor that interested me very much—namely, that there might be a sorcerer on this island. Now, sorcerers have fascinated me since childhood, so I was very anxious to meet a real one, if the rumor was true. I wondered how I could find out!

14

Fortunately, a native asked me a question one day, and then I knew that there must be a sorcerer on the island.

Can you supply such a question?

At this point, the reader might well be wondering how I could possibly have heard a rumor about a sorcerer on the island or, for that matter, heard anything at all about the island, since the inhabitants never make statements but only ask questions. Assuming that the reader hasn't already figured out the answer for himself, the solution to this problem will show exactly how the inhabitants can communicate information just as freely (if somewhat more clumsily) as anyone else.

As you can imagine, I was delighted to find out that there really was a sorcerer on the island. I also learned that he was the island's only sorcerer. But I had no idea who he was. Then I discovered that a grand prize had been offered to any visitor who could correctly guess his name. The only drawback was that any visitor who guessed wrong would be executed.

So I got up early the next morning and walked around the island, hoping that the natives would ask me enough questions to enable me to deduce with certainty who the sorcerer was. Here is what happened:

15

The first native I met was named Arthur Good. He asked me, "Am I the sorcerer?"

Did I have enough information yet to know who the sorcerer was?

16

The next native was named Bernard Green. He asked me, "Am I the type who could ask whether I am not the sorcerer?"

Did I yet have enough information?

17

The next native, Charles Mansfield, asked, "Am I the type who could ask whether the sorcerer is the type who could ask whether I am the sorcerer?"

Do I yet have enough information?

18

The next native was named Daniel Mott. He asked, "Is the sorcerer of type B?"

Do I yet have enough information?

19

The next native was named Edwin Drood. He asked, "Are the sorcerer and I of the same type?"

Eureka! I now had enough pieces to solve the mystery!

Who is the sorcerer?

◇ ◇ ◇

Bonus Problem

Are you a good detective? We recall the patient Thomas who visited this island. Is he really sane or insane?

• SOLUTIONS •

1 • It is impossible for any native of this island to ask you this question. If a native of type A asks, "Am I of type B?," the correct answer is no (since he isn't of type B), but a type A cannot ask any question whose correct answer is no. Therefore, no native of type A can ask this question. If a native of type B asks the question, the correct answer is yes (since he *is* of type B), but a type B cannot ask a question whose correct answer is yes. Therefore, a native of type B cannot ask the question either.

2 • Nothing can be concluded. Any native of this island can ask whether he is of type A, because he is of either type A or type B. If he is of type A, then the correct answer to the question, "Am I of type A?" is yes, and anyone of type A can ask any question whose correct answer is yes. On the other hand, if the inhabitant is of type B, then the correct answer to the question is no, and any inhabitant of type B can ask a question whose correct answer is no.

3 ◆ We must first find out Ethan's type. Suppose Ethan is of type A. Then the correct answer to his question must be yes (since yes is the correct answer to questions asked by those of type A), which would mean that Ethan and Violet are both of type B, which would mean that Ethan is of type B, and we have a contradiction. Therefore, Ethan can't be of type A; he must be of type B. Since he is of type B, the correct answer to his question is no, so it is not the case that he and Violet are both of type B. This means Violet must be of type A.

4 ◆ Suppose Arthur were of type B. Then it would be true that at least one of the brothers was of type B, which would make yes the correct answer to his question, which would mean he is of type A. This is a contradiction; hence Arthur cannot be of type B; he must be of type A. From this it follows that the correct answer to his question is yes, which means that at least one of the two is of type B. Since Arthur is not of type B, this must be Robert. So Arthur is of type A and Robert is of type B.

5 ◆ Nothing can be deduced about Mr. Gordon, but Mrs. Gordon must be of type B. Here are the reasons why:

Mr. Gordon is of either type A or type B. Suppose he is of type A. Then the correct answer to his question is yes, which means the two are of different types. This means Mrs. Gordon must be of type B (since he is of type A and the two are of different types). So, if Mr. Gordon is of type A, then Mrs. Gordon must be of type B.

Now, suppose Mr. Gordon is of type B. Then the correct

answer to his question is no, which means the two are not of different types; they are of the same type. This means that Mrs. Gordon is also of type B. So if Mr. Gordon is of type B, so is Mrs. Gordon.

This proves that regardless of whether Mr. Gordon is of type A or type B, Mrs. Gordon must be of type B.

Another proof—much simpler but more sophisticated—is this: We already know from the first problem that no one on this island can ask whether he is of type B. Now, if Mrs. Gordon were of type A, then for an inhabitant to ask whether he is of a different type from Mrs. Gordon would be equivalent to his asking whether he is of type B, which he cannot do. Therefore, Mrs. Gordon cannot be of type A.

6 ◆ This story is perfectly possible, but Zorn must be of type B. The easiest way to see this is by recalling (Problem 1) that no native of this island can ask whether he is of type B. So when Zorn asks whether he is the type who could ask whether he is of type B, the correct answer is no (since no inhabitant could ask whether he is of type B). Since the correct answer is no, Zorn must be of type B.

7 ◆ Since the native just *did* ask the question, then he obviously *could* ask the question. Hence the correct answer to his question is yes, and he is of type A.

8 ◆ Nothing can be deduced about Mrs. Klink, but Mr. Klink must be of type A. Here are the reasons: Suppose Mrs. Klink is of type A. Then the correct answer to her question is yes, which means that Mr. Klink *could* ask Mrs. Klink if she is of type A. And, since Mrs. Klink is of type A, the correct answer would be *yes,* which makes Mr. Klink of type A. So, if Mrs. Klink is of type A, so is her husband. Now, suppose Mrs.

Klink is of type B. Then the correct answer to her question is no, which means that Mr. Klink is not the type who could ask her if she is of type A. Thus he could not ask a question whose correct answer is no, so he must be of type A. So, Mr. Klink is of type A, regardless of the type of Mrs. Klink.

9 ◆ Suppose Betty is of type A. Then the correct answer to her question is yes; hence John could ask if at least one of them is of type B. But this leads to a contradiction: If John is of type A, then it would be false that at least one of them is of type B. Hence the correct answer to his question would be no, which is not possible for one of type A. If John is of type B, then it would be true that at least one of them is of type B, which makes yes the correct answer to his question. But one of type B cannot ask a question whose correct answer is yes. Thus the assumption that Betty is of type A is impossible; she must be of type B.

Since Betty is of type B, then the correct answer to her question is no, which means that John cannot ask her if at least one of them is of type B. Now, if John were of type A, then he *could* ask that question, because it is true that at least one of them is of type B (namely, Betty). Since he can't ask that question, he must also be of type B.

So the answer is that both of them are of type B.

10 ◆ It is easiest to build the solution of this problem in graded steps. First, we can easily establish the following two propositions:

Proposition 1: Given any inhabitant X of type A, no inhabitant could ask whether (s)he and X are of different types.

Proposition 2: Given any inhabitant X of type B, then any inhabitant could ask whether (s)he and X are of different types.

We have already proved Proposition 1 in the solution of Problem 5, in which we saw that if Mrs. Gordon had been of type A, then Mr. Gordon *couldn't* have asked whether he and Mrs. Gordon were of the same type.

As for Proposition 2, if X is of type B, then the question of whether one is of a different type than X is equivalent to the question of whether one is of type A, and anyone can ask that question, as we saw in the solution of Problem 2. Therefore, anyone can ask X whether (s)he is of a different type, if X is of type B.

Now for the problem: I will prove that the correct answer to Alice's question is no; hence Alice must be of type B. In other words, I will prove that it is *not* possible for Betty to ask Cynthia whether Cynthia is the type who could ask Betty whether Cynthia and Betty are of different types.

Suppose Betty asks Cynthia whether Cynthia could ask whether Cynthia and Betty were of different types. We get the following contradiction: Betty is either of type A or type B. Suppose Betty is of type A. Then by Proposition 1, Cynthia could not ask whether she and Betty are of different types; hence the answer to Betty's question is no, which is impossible since Betty is of type A! On the other hand, suppose Betty is of type B. Then by Proposition 2, Cynthia *could* ask whether she and Betty are of different types, which makes yes the correct answer to Betty's question, which is not possible since Betty is of type B.

This proves that Betty could never ask Cynthia the question which Alice asks Betty whether she could ask, so the correct answer to Alice's question is no, and Alice is of type B. As to the types of Betty and Cynthia, nothing can be determined.

11 ◆ This strikes me as the funniest problem of this chapter, since nothing can be deduced about the native who asked the question; but as to Arnold, though he never opened his mouth (as far as we know), he must be insane! The fact is that no native could ask a *sane* person whether he believes the native to be of type B, because asking a sane person whether he believes such-and-such to be the case is tantamount to asking whether such-and-such really is the case, and no native can ask whether he is of type B. So no native X could ask a sane person whether he believes X is of type B.

On the other hand (and we need this fact for a subsequent problem), any native X could ask an *insane* person whether he believes X is of type B, because asking that of an insane person is tantamount to X asking whether X is of type A, which, as we have seen, any native X can do.

12 ◆ Nothing can be deduced about Thomas, but the native who asked the question must be of type B. For, suppose he were of type A, then the correct answer to his question is yes, which means that Thomas does believe that the native could ask him whether he is insane. Now, Thomas is either sane or insane. Suppose he is sane. Then his belief is correct, which means that the native could ask him if he is insane. But one of type A can ask a question only if the correct answer is yes, which would mean that Thomas must be insane; so the assumption that Thomas is sane leads to the conclusion that Thomas is insane. Therefore, it is contradictory to assume that Thomas is sane. On the other hand, suppose Thomas is insane. Then Thomas's belief that the native could ask if Thomas is insane is wrong; hence the native couldn't ask him if he is insane. (Thomas would answer no—impossible, given that the native is of type A.) But, given that Thomas is insane, and the native is of type A, then the native *could,* by the rules on the Island of Questioners, ask Thomas if he is insane (since the correct answer would be yes). So it is also contradictory to assume that Thomas is insane.

The only way out of the contradiction is that the native must be of type B rather than type A, and no contradiction arises, regardless of whether Thomas is sane or insane.

13 ◆ I will show that the story which William reported could never really have happened; hence William must be insane to believe that it had.

Suppose the story were true: we get the following contradiction. Suppose Thomas is sane; then his statement is correct, which means that Hal could ask Thomas if he believes that Hal is of type B. But by the solution to Problem 11, this implies that Thomas is insane! So it is contradictory to assume that Thomas is sane. On the other hand, suppose Thomas is insane. Then his statement is false; hence Hal couldn't ask Thomas if he believes that Hal is of type B. But, as we saw in Problem 11, a native *can* ask an insane person whether he believes the native to be of type B; so we also get a contradiction in this case.

The only way out of the contradiction is that Thomas never did ask such a question of any native, and William only imagined he did.

14 ◆ Many questions will do the trick; my favorite one is: "Am I the type who can ask whether there is a sorcerer on this island?"

Suppose the questioner is of type A. Then the correct answer to his question is yes; the questioner *can* ask if there is a sorcerer on the island. Being of type A, he can ask whether there is a sorcerer on the island only if there is in fact a sorcerer on the island (so the correct answer would be yes). Thus, if the questioner is of type A, then there must be a sorcerer on the island.

Suppose the questioner is of type B. Then the correct answer to his question is no, which means that he cannot ask if there is a sorcerer on the island. Now if there *were* no sorcerer on the island, then the questioner (being of type B) *could* ask if there is a sorcerer on the island (since the correct answer would be no). However, since the questioner can't ask this (as we have seen), it follows that there must in fact be a sorcerer on the island. This proves that if the questioner is of type B, there is a sorcerer on the island. So, regardless of whether the questioner is of type A or type B, there must be a sorcerer on the island.

15 ◆ Of course not!

16 ◆ All that can be deduced is that Bernard Green is not the sorcerer (by the same reasoning as in the solution of Problem 14).

17 ◆ All that can be deduced is that the sorcerer is the type who could ask if Charles Mansfield is the sorcerer. (Remember that, as we found out in Problem 11, when a native asks, "Am I the type who could ask such-and-such?" then the such-and-such must in fact be true.)

18 ◆ All that can be deduced is that Daniel Mott is not the sorcerer (because the sorcerer cannot ask whether the sorcerer is of type B; no one can ask whether he is of type B).

19 ◆ It is not possible to deduce who the sorcerer is from what Edwin Drood asks by itself, but from Edwin Drood's question *together with earlier questions,* the problem becomes completely resolved!

What follows from Edwin Drood's question is that the sorcerer must be of type A. For, suppose Edwin is of type A. Then the correct answer to his question is yes; hence he and the sorcerer really would be of the same type; so the sorcerer would also be of type A. On the other hand, suppose Edwin is of type B. Then the correct answer to his question is no, which means that he and the sorcerer would not be of the same type. Since Edwin would be of type B, and the sorcerer would not be of the same type as Edwin, then again the sorcerer must be of type A.

This proves that the sorcerer is of type A. Now, we saw in Problem 17 that the sorcerer could ask if Charles Mansfield is the sorcerer. Since the sorcerer is of type A, then the correct answer to that question is yes; hence Charles Mansfield must be the sorcerer!

Bonus Problem ◆ I told you that Arnold, Thomas, and William were unanimously agreed that life outside the asylum was even crazier than life inside. Since Thomas agrees with Arnold and William, who are insane, then Thomas must also be insane.

◆ 6 ◆

The Isle of Dreams

I once dreamed that there was a certain island called the Isle of Dreams. The inhabitants of this island dream quite vividly; indeed, their thoughts while asleep are as vivid as while awake. Moreover, their dream life has the same continuity from night to night as their waking life has from day to day. As a result, some of the inhabitants sometimes have difficulty in knowing whether they are awake or asleep at a given time.

Now, it so happens that each inhabitant is of one of two types: *diurnal* or *nocturnal.* A diurnal inhabitant is characterized by the fact that everything he believes while he is awake is true, and everything he believes while he is asleep is false. A nocturnal inhabitant is the opposite: everything a nocturnal person believes while asleep is true, and everything he believes while awake is false.

1

At one particular time, one of the inhabitants believed that he was of the diurnal type.

Can it be determined whether his belief was correct? Can it be determined whether he was awake or asleep at the time?

2

On another occasion, one of the natives believed he was asleep at the time. Can it be determined whether his belief was correct? Can it be determined what type he is?

3

(a) Is it true that an inhabitant's opinion of whether he is diurnal or nocturnal never changes?

(b) Is it true that an inhabitant's opinion of whether he is awake or asleep at the time never changes?

4

At one time, an inhabitant believed that she was either asleep or of the nocturnal type, or both. (*Or* means *at least one* or *possibly both.*)

Can it be determined whether she was awake or asleep at the time? Can it be determined what type she is?

5

At one time, an inhabitant believed that he was both asleep and diurnal. What was he really?

6

There is a married couple on this island whose last name is Kulp. At one point Mr. Kulp believed that he and his wife were both nocturnal. At the same instant, Mrs. Kulp believed that they were not both nocturnal. As it happened, one of them was awake and one of them was asleep at the time. Which one of them was awake?

7

There is another married couple on this isle whose last name is Byron. One of them is nocturnal and the other is diurnal. At one point the wife believed that they were either both asleep or both awake. At the same instant, the husband believed that they were neither both asleep nor both awake.

Which one was right?

8

Here is a particularly interesting case: at one time an inhabitant named Edward believed amazingly that he and his sister Elaine were both nocturnal, and at the same time that he was not nocturnal.

How is this possible? Is he nocturnal or diurnal? What about his sister? Was he awake or asleep at the time?

9 • The Royal Family

This isle has a king and a queen and also a princess. At one point the princess believed that her parents were of different types. Twelve hours later, she changed her state (either from sleeping to waking or from waking to sleeping), and she then believed that her father was diurnal and her mother was nocturnal.

What type is the king and what type is the queen?

10 • And What About the Witch Doctor?

No island is complete without a sorcerer or magician, a medicine man or witch doctor, or something like that. Well, this island, as it happens, has a witch doctor and only one witch doctor. Now comes a particularly intriguing puzzle concerning this witch doctor:

At one time an inhabitant named Ork was wondering whether he himself was the witch doctor. He came to the conclusion that if he was diurnal and awake at that point, then he must be the witch doctor. At the same instant, another inhabitant named Bork believed that if he was either diurnal and awake or nocturnal and asleep, then he (Bork) was the witch doctor. As it happened, Ork and Bork were either both asleep or both awake at the time.

Is the witch doctor diurnal or nocturnal?

11 • A Metapuzzle

I once gave a friend the following puzzle about this island:

"An inhabitant believed at one time that he was diurnal and awake. What was he really?"

My friend thought about this for a while and then replied, "You obviously haven't given me enough information!" Of course my friend was right! He then asked me, "Do *you* know what type he was and whether he was awake or asleep at the time?"

"Oh, yes," I replied, "I happen to know this inhabitant well, and I know both his type and his state at the time."

My friend then asked me a shrewd question: "If you were to tell me whether he was diurnal or nocturnal, would I then have enough information to know whether he was awake or asleep at the time?" I answered him truthfully (yes or no), and he was then able to solve the puzzle.

Was the inhabitant diurnal or nocturnal, and was he awake or asleep at the time?

12 • A More Difficult Metapuzzle

On another occasion, I told a friend the following puzzle concerning this island:

"An inhabitant at one point believed that she was both asleep and nocturnal. What was she really?"

My friend immediately realized that I had not given him enough information.

"Suppose you told me whether the lady was nocturnal or diurnal," my friend asked me. "Would I then be able to deduce whether she was asleep or awake at the time?"

I answered him truthfully, but he was *not* able to solve the problem (he still hadn't enough information).

Some days later, I gave the same problem to another friend (without telling him about the first friend). This second friend also realized that I hadn't given him enough information. Then he asked me the following question: "Suppose you told me whether the lady was awake or asleep at the time; would I then have enough information to know whether she was diurnal or nocturnal?"

I answered him truthfully, but he was unable to solve the problem (he too did not have enough information).

At this point, *you* have enough information to solve the puzzle! Was the lady diurnal or nocturnal, and was she awake or asleep at the time?

Epilogue

Suppose there really existed an island of the type described in this chapter, and suppose that I were one of the inhabitants. Would I be the diurnal or nocturnal type? It is really possible to answer this on the basis of things I have said in this chapter!

• SOLUTIONS •

1, 2, 3 • Let us first observe that the following laws must hold:

Law 1: An inhabitant while awake believes he is diurnal.

Law 2: An inhabitant while asleep believes he is nocturnal.

Law 3: Diurnal inhabitants at all times believe they are awake.

Law 4: Nocturnal inhabitants at all times believe they are asleep.

To prove Law 1: Suppose X is an inhabitant who is awake at a given time. If X is diurnal, then he is both diurnal and awake; hence his beliefs at the time are correct; and he knows he is diurnal. On the other hand, suppose X is nocturnal. Then, being nocturnal but awake at the time, his beliefs are wrong; hence he erroneously believes he is diurnal. In summary, if X is awake, then if he is diurnal, he (rightly) believes he is diurnal, and if he is nocturnal, he (wrongly) believes he is diurnal.

The proof of Law 2 is parallel: If X is asleep, then if he is nocturnal, he (rightly) believes he is nocturnal, and if he is diurnal, he (wrongly) believes he is nocturnal.

To prove Law 3, suppose X is diurnal. While awake, his beliefs are correct; hence he then knows he is awake. But while asleep, his beliefs are wrong; hence he then erroneously believes he is awake. So, while awake he (rightly) believes he is awake, and while asleep he (wrongly) believes he is awake.

The proof of Law 4 parallels the proof of Law 3, and is left to the reader.

Now, to solve Problem 1, it cannot be determined whether his belief was correct. But he must have been awake at the time, for had he been asleep, he would have believed himself nocturnal rather than diurnal (by Law 2).

As for Problem 2, again it cannot be determined if his belief was correct; but the native must have been nocturnal, for were he diurnal, he would have believed himself to be awake rather than asleep (by Law 3).

As for Problem 3, the answer to (a) is no (because by Laws 1 and 2, an inhabitant's opinion as to whether he is diurnal or nocturnal changes from state to state; that is, from the waking state to the sleeping state), but the answer to (b) is yes (by Laws 3 and 4).

4 ◆ You can solve this systematically by considering each of the four possibilities in turn: (1) she is nocturnal and asleep; (2) she is nocturnal and awake; (3) she is diurnal and asleep; (4) she is diurnal and awake. You can then see which of the possibilities is compatible with the given conditions. However, I prefer the following argument:

First of all, could her belief be incorrect? If it is, then she is neither asleep nor nocturnal, which means she is awake and diurnal. However, this is a contradiction, since a person who is awake and diurnal cannot have an incorrect belief. Therefore, her belief cannot be incorrect; it must be correct. This means that she is asleep and nocturnal.

5 ◆ Again, this could be solved by trying each of four answers in turn, but again I prefer a more creative solution.

Could his belief have been correct? If so, then he was really asleep and diurnal, but being asleep and diurnal, he couldn't have a correct belief. Therefore, his belief was wrong. Now, the only occasions on which an inhabitant can have a wrong belief is when he is either asleep and diurnal or awake and nocturnal. If he were asleep and diurnal, then his belief would have been correct (for that is what he believes). Hence he must have been awake and nocturnal.

6 ◆ If you go about solving this puzzle systematically, you will have sixteen cases to consider! (Four possibilities for the husband, and with each of these four possibilities there are four possibilities for the wife.) Fortunately, there is a much simpler method of approaching the problem.

To begin with, since one of the two is asleep and the other awake, and since they believe opposite things, then they must be of the same type (that is, either both diurnal or both nocturnal), because if they were of different types, their beliefs would be opposite when they were both asleep or both awake and would coincide when one was asleep and the other was awake. Since their beliefs when one is asleep and the other awake don't coincide, then they must be of the same type.

Given, therefore, that they are either both nocturnal or both diurnal, let us suppose they are both nocturnal. Then the husband's belief at the time was correct, and since he is nocturnal, he must have been asleep at the time. Now, suppose they are both diurnal. Then the husband was obviously wrong in believing that they were both nocturnal, and since he is diurnal and had a wrong belief, then he must have been asleep at the time. So, whether they are nocturnal or diurnal, the husband must have been asleep at the time and his wife awake.

7 ◆ This is even simpler: Since the husband and wife are of different types, then their beliefs must be opposite when they

are in the same state (that is, both awake or both asleep), and their beliefs must be the same when they are in different states (one asleep and the other awake). Since on this occasion their beliefs were opposite, then they were in the same state—both asleep or both awake. Therefore, the wife was right.

8 ◆ Obviously, Edward must have been in an unreliable state of mind at the time to believe these two logically incompatible propositions! So, both of Edward's beliefs must be wrong. Since he believed that he and Elaine were both nocturnal, then they are not both nocturnal. And since he believed he was not nocturnal, then he *is* nocturnal. So he is nocturnal, but they are not both nocturnal, so Elaine is diurnal. Since he is nocturnal and believed falsely at the time, he must have been awake. So, the answer is that he is nocturnal, his sister is diurnal, and he was awake.

9 ◆ Since the princess changed her state, then one of her two beliefs was correct and the other incorrect. This means that of the following two propositions, one is true and the other is false:

(1) The king and queen are of different types.

(2) The king is diurnal and the queen is nocturnal.

If (2) is true, then (1) would also have to be true, but we know that (2) and (1) can't both be true. Therefore, (2) must be false, and hence also (1) must be true. So the king and queen really are of different types, but it is not the case that the king is diurnal and the queen nocturnal. Therefore, the king is nocturnal and the queen is diurnal.

10 ◆ Suppose Ork were diurnal and awake at the time; would it follow from that supposition that Ork must be the witch doctor? Yes it would, by the following argument: Suppose Ork really were diurnal and awake at the time. Then his belief is correct, which means that *if* he is diurnal and awake, then he *is* the witch doctor. But he is diurnal and awake (by supposition); hence he must be the witch doctor (still under the supposition, of course, that he is diurnal and awake). So, the supposition that he is diurnal and awake leads to the conclusion that he is the witch doctor. This, of course, does not prove that the supposition is true, nor that he is the witch doctor, but only that *if* he was diurnal and awake, then he is the witch doctor. So we have established the hypothetical proposition that *if* Ork was diurnal and awake, *then* he is the witch doctor. Well, it was precisely this hypothetical proposition which Ork believed at the time; therefore, Ork's belief was correct! This means that Ork was either diurnal and awake at the time, or nocturnal and asleep, at the time, but we cannot (yet) tell which. Therefore, it is not necessarily true that Ork is the witch doctor, since it could be that he was nocturnal and asleep at the time.

Now, by a rather similar argument, Bork's belief is also correct: If Bork is either diurnal and awake or nocturnal and asleep, *in either case,* his belief is correct, which means he would have to be the witch doctor. Well, this is precisely what Bork believes; so Bork's belief is correct. Since Bork's belief is correct, then either he is diurnal and was awake at the time, or he is nocturnal and was asleep at the time. But in either case he must be the witch doctor.

Since Bork is the witch doctor, then Ork is not. Therefore, Ork could not have been awake at the time and diurnal, for we showed that if he had been, then *he* would have been the witch doctor. So, Ork was asleep at the time, and also nocturnal. Therefore, Bork was also asleep at the time, and since Bork's belief at the time was correct, then Bork must be nocturnal. So the witch doctor is nocturnal.

11 ◆ From the fact that the inhabitant believed that he was diurnal and awake, all that follows is that he was not nocturnal and asleep, and so there are three possibilities:

(1) He was nocturnal and awake (and believed falsely).

(2) He was diurnal and asleep (and believed falsely).

(3) He was diurnal and awake (and believed truly).

Now, suppose I had told my friend whether the native was diurnal or nocturnal; could my friend have then solved the problem? Well, that would depend on what I told him. If I told him that the native was nocturnal, then he would have known that Case 1 above was the only possibility, and so he would have known that the native was awake. On the other hand, if I told him that the native was diurnal, that would have ruled out (1) but would leave open both (2) and (3), and my friend wouldn't have any way of knowing which of these two latter possibilities actually held; so he then could not have solved the problem.

Now, my friend did not ask me whether the native was diurnal or nocturnal; all he asked was whether he could solve the problem *if* I told him whether the native was diurnal or nocturnal. If, in fact, the native were diurnal, then I would have had to answer no to my friend's question (because, as I have shown, if I told him that the native was diurnal, he couldn't solve the problem), but if the native were nocturnal, then I would have had to answer yes to his question (because, as I have shown, if I told him that the native was nocturnal, then my friend could solve the problem). Therefore, since my friend knew that the native was nocturnal and awake, I must have answered yes.

12 ◆ From the fact that she believed that she was nocturnal and asleep, all that follows is that she was not diurnal and awake, and so three possibilities remain:

(1) She was nocturnal and asleep.
(2) She was nocturnal and awake.
(3) She was diurnal and asleep.

If I had answered yes to my first friend's question, he would have known that (3) is the only possibility (by an argument similar to the solution of the last puzzle). But since he didn't solve it, I must have answered no. This, then, rules out (3), so we are left with possibilities (1) and (2).

Now, consider my second friend. If I had answered yes, then he could have figured out that (2) is the only real possibility (because (2) is the only one in which she is awake, whereas (1) and (3) both hold if she is asleep). Since this second friend couldn't solve the problem either, I must have answered him no as well, and this rules out possibility (2). What remains is that possibility (1) is the only valid one—that is, the native was nocturnal and asleep, as she herself correctly believed.

In brief summary, the fact that my first friend couldn't solve the problem rules out (3), and the fact that my second friend couldn't solve it rules out (2). What remains is (1): she was nocturnal and asleep.

Epilogue ◆ I told you at the beginning of the chapter that I *dreamed* there was such an island. If there really were such an island, then I would have dreamed truly; hence if I were one of the inhabitants, I would have to be nocturnal.

◆ 7 ◆ Metapuzzles

The last two puzzles of the last chapter (not counting the epilogue) are examples of a fascinating type of puzzle that I am tempted to call *metapuzzles*—or puzzles about puzzles. We are given a puzzle without sufficient data to solve it, and then we are given that someone else could or could not solve it given certain additional information, but we are not always told just what this additional information is. We may, however, be given partial information about it, which enables the reader to solve the problem. This remarkable genre is unfortunately rather rare in the literature. What follow here are five such puzzles, starting with some very easy ones and progressing to the last, which is the crowning puzzle of this and the preceding chapters.

1 ◆ The Case of John

This case involved a judicial investigation of identical twins. It was known that at least one of them never told the truth, but it was not known which. One of the twins was named John, and he had committed a crime. (John was not necessarily the one who always lied.) The purpose of the investigation was to find out which one was John.

"Are you John?" the judge asked the first twin.

"Yes, I am," was the reply.

"Are you John?" the judge asked the second twin.

The second twin then answered either yes or no, and the judge then knew which one was John.

Was John the first twin or the second?

2 • A Transylvanian Metapuzzle

We learned from Chapter 4 that every Transylvanian is one of four types: (1) a sane human; (2) an insane human; (3) a sane vampire; (4) an insane vampire. Sane humans make only true statements (they are both accurate and honest); insane humans make only false statements (out of delusion, not intention); sane vampires make only false statements (out of dishonesty, not delusion); and insane vampires make only true statements (they believe the statement is false, but lie and say the statement is true).

Three logicians were once discussing their separate trips to Transylvania.

"When I was there," said the first logician, "I met a Transylvanian named Igor. I asked him whether he was a sane human. Igor answered me [yes or no], but I couldn't tell from his answer what he was."

"That's a surprising coincidence," said the second logician. "I met that same Igor on *my* visit. I asked him whether he was a sane vampire and he answered me [yes or no], and I couldn't figure out what he was."

"This is a double coincidence!" exclaimed the third logician. "I also met Igor and asked him whether he was an insane vampire. He answered me [yes or no], but I couldn't deduce what he was either."

Is Igor sane or insane? Is he a human or a vampire?

3 • A Knight-Knave Metapuzzle

My book *What Is the Name of This Book?* contains a host of puzzles about an island on which every inhabitant is either a knight or a knave; knights always tell the truth and knaves always lie. Here is a knight-knave metapuzzle.

A logician once visited this island and came across two inhabitants, A and B. He asked A, "Are both of you knights?" A answered either yes or no. The logician thought for a while, but did not yet have enough information to determine what they were. The logician then asked A, "Are you two of the same type?" (Same type means both knights or both knaves.) A answered either yes or no, and the logician then knew what type each one was.

What type is each?

4 • Knights, Knaves, and Normals

On the island of Knights, Knaves, and Normals, knights always tell the truth, knaves always lie, and those called *normal* can either lie or tell the truth (and sometimes one and sometimes the other).

One day I visited this island and met two inhabitants, A and B. I already knew that one of them was a knight and the other was normal, but I didn't know which was which. I asked A whether B was normal, and he answered me, either yes or no. I then knew which was which.

Which of the two is normal?

5 • Who Is the Spy?

Now we come to a far more intricate metapuzzle!

This case involves a trial of three defendants: A, B, and C. It was known at the outset of the trial that one of the three was a knight (he always told the truth), one a knave (he always lied), and the other was a *spy* who was normal (he sometimes lied and sometimes told the truth). The purpose of the trial was to find the spy.

First, A was asked to make a statement. He said either that C is a knave or that C was the spy, but we are not told which. Then B said either that A is a knight, or that A is a knave, or that A was the spy, but we are not told which. Then C made a statement about B, and he said either that B was a knight, or that B was a knave, or that B was the spy, but we are not told which. The judge then knew who the spy was and convicted him.

This case was described to a logician, who worked on the problem for a while, and then said, "I do not have enough information to know which one is the spy." The logician was then told what A said, and he then figured out who the spy was.

Which one is the spy—A, B, or C?

• SOLUTIONS •

1 • If the second twin had also answered yes, the judge obviously could not have known which one was John; hence the second one must have answered no. This means that either both twins told the truth or both lied. But they couldn't have both told the truth, because it is given that at least one of

them always lies. Therefore, both lied, which means that the second twin is John. (It cannot be decided which of the two always lies.)

2 ◆ The first logician asked Igor whether he was a sane human. If Igor *is* a sane human, he would answer yes; if he is an insane human, he would also answer yes (because, being insane, he would erroneously believe that he is a sane human and then honestly express this belief); if Igor is a sane vampire, he would also answer yes (because, being sane, he knows he isn't a sane human, but would lie and say he was); but if Igor is an insane vampire, then he would answer no (because, being an insane vampire, he believes he is a sane human and lies about what he believes). So an insane vampire will answer no to this question; the other three types will answer yes. Now, if Igor had answered no, then the first logician would have known that Igor was an insane vampire. But the first logician didn't know what Igor was; hence he must have gotten a yes answer. All we can infer from this is that Igor is not an insane vampire.

As to the second logician's question, "Are you a sane vampire?," an insane human would answer yes, and each of the other three types would answer no. (We leave the verification of this to the reader.) Since the second logician couldn't tell from Igor's answer what Igor was, the answer must have been no, which means that Igor is not an insane human.

As to the third logician's question, "Are you an insane vampire?," a sane human would answer no, and each of the other three types would answer yes. Since the third logician couldn't figure out what Igor was, he must have gotten the answer yes, which means that Igor is not a sane human.

Since Igor is neither an insane vampire nor an insane human nor a sane human, he must be a sane vampire.

3 ◆ There are four possible cases:

Case 1: A and B are both knights.

Case 2: A is a knight and B is a knave.

Case 3: A is a knave and B is a knight.

Case 4: A and B are both knaves.

The logician first asked A whether both of them were knights. If Case 1, Case 3, or Case 4 holds, A will answer yes; if Case 2 holds, A will answer no. (We leave the verification of this to the reader.) Since the logician didn't know from A's answer what the natives were, then A must have answered yes. All the logician then knew was that Case 2 was out. Next, the logician asked A whether both were of the same type. In Cases 1 and 3, A would answer yes, and in Cases 2 and 4, A would answer no. (Again, I leave the verification of this to the reader.) So if the logician had gotten the answer yes, all he would have known is that either Case 1 or Case 3 holds, but he wouldn't know which. So he must have gotten the answer no. He then knew that either Case 2 or Case 4 holds, but he had already ruled out Case 2. So he knew that Case 4 must hold. And so A and B are both knaves.

4 ◆ If A had replied yes, then A could have been a knight, or A could have been normal (and lied), and I couldn't have known which. If A had replied no, then A couldn't be a knight (for then B would be normal, and A would have lied); so A would have to be normal. The only way I could have known which was which is that A said no. Hence A is the normal one.

5 ◆ We, of course, assume that the judge was a perfect reasoner and also that the logician to whom the problem was told was a perfect reasoner.

There are two possibilities: either the logician was told

that A said that C was a knave, or he was told that A said that C is the spy. We must examine both possibilities.

Possibility I: A said that C is a knave.

There are now three possible cases for what B said, and we must examine each:

Case 1: B said that A is a knight. Then: (1) if A is a knight, C is a knave (because A said that C is a knave), hence B is the spy; (2) if A is a knave, then B's statement is false, which means that B must be the spy (he's not a knave since A is), hence C is a knight; (3) if A is the spy, then B's statement is false, which means B is the knave, hence C is the knight. Thus we have either:

(1) A knight, B spy, C knave.

(2) A knave, B spy, C knight.

(3) A spy, B knave, C knight.

Now, suppose C said that B is the spy. Then (1) and (3) are ruled out. (If (1), C, a knave, couldn't claim that B is a spy, because B is, and if (3), C, a knight, couldn't claim that B is a spy, because B isn't.) This leaves only (2) open, and the judge would then know that B was the spy.

Suppose C said that B is a knight. Then (1) is the only possibility, and the judge would know this and again convict B.

Suppose C said that B is a knave. Then the judge couldn't have known whether (1) or (3) holds; hence he couldn't have known whether A or B was the spy, so he couldn't have convicted anyone. Therefore, C didn't say that B is a knave. (Of course, we are still working under the assumption for Case 1—that B said that A is a knight.)

So, if Case 1 holds, then B is the only one the judge could have convicted.

Case 2: B said that A is the spy. We leave it to the reader to verify that the following are the only possibilities:

(1) A knight, B spy, C knave.

(2) A knave, B spy, C knight.

(3) A spy, B knight, C knave.

If C said that B is the spy, then either (2) or (3) could hold, and the judge couldn't have found anyone guilty. If C said that B is a knight, then only (1) can hold, and the judge convicted B. If C said that B is a knave, then either (1) or (3) could hold, and the judge couldn't have convicted anyone. Therefore, C must have said that B is a knight, and B was the one convicted.

So, under Case 2, B was again the one convicted.

Case 3: B said that A is a knave. In this case there are four possibilities (as the reader can verify):

(1) A knight, B spy, C knave.

(2) A knave, B spy, C knight.

(3) A knave, B knight, C spy.

(4) A spy, B knave, C knight.

If C said that B is the spy, (2) or (3) could hold, and the judge couldn't have determined which one was guilty. If C said that B is a knight, (1) or (3) could hold, and the judge, again, couldn't have convicted anyone. If C said that B is a knave, (1), (3), or (4) could hold, and once more the judge could not have determined where guilt lay.

Thus Case 3 is ruled out. So we now know that either Case 1 or Case 2 holds, and in both cases, the judge convicted B.

So if Possibility I obtains (if A said that C is a knave), then B must be the spy. Therefore, if the logician had been told that A said that C is a knave, he could solve the problem and know that B was the spy.

Possibility II: Now, suppose the logician had been told that A said that C is the spy. I will show that the logician would then be unable to solve the problem, because there would be a possibility that the judge convicted A and a possibility that the judge convicted B, and the logician couldn't know which.

To prove this, let us assume that A said that C was the spy. Then here is one way the judge could have convicted A: Suppose B said that A is a knight and C said that B is a knave. If A is the spy, B could be a knave (who falsely claimed that A is a knight), and C could be a knight (who truthfully claims that B is a knave). A (the spy) would have falsely claimed that C is the spy. So it really is possible that A, B, and C made these three statements and that A is the spy. Now, if B were the spy, then A would have to be a knave in order to claim that C is the spy, and C would also have to be a knave for claiming that B is a knave, and so this is not possible. If C were the spy, then A would have to be a knight for truthfully claiming that C is a spy, and B would also have to be a knight for truthfully claiming that A is a knight, so this is also not possible. Therefore, A must be the spy (if B said that A is a knight and C said that B is a knave). So it is possible that A was the one convicted.

Here is a way that B could have been convicted: suppose B said that A is a knight and C said that B is the spy. (We continue to assume that A said that C is the spy.) If A is the spy, B is a knave for saying that A is a knight and C is also a knave for saying that B is the spy, so this is not possible. If C is the spy, then A is a knight (since he said C is the spy), and B is also a knight for saying that A is a knight, so this is also not possible. But if B is the spy, there is no contradiction (A could be a knave who said C is the spy; C could be a knight who said B is the spy; and B could have said that A is a knight). So it is possible that A, B, and C did make these three statements, in which case the judge convicted B.

I have now shown that if A said that C was the spy, there is a possibility that the judge convicted A and a possibility that the judge convicted B, and there is no way to tell which. Therefore, if the logician had been told that A said that C was the spy, there is no way the logician could have solved the problem. But we are given that the logician did solve the problem; hence he must have been told that A said that C is a knave. Then (as we have seen), the judge could have convicted only B. So B is the spy.

PART THREE

THE MYSTERY OF THE MONTE CARLO LOCK

◆ 8 ◆

The Mystery of the Monte Carlo Lock

We last left Inspector Craig seated comfortably aboard a train outward bound from Transylvania, relieved at the thought of returning home. "Enough of these vampires!" he said to himself. "I'll be glad to get back to London, where things are normal!"

Little did Craig realize that another adventure awaited him before his return—an adventure of a very different nature from the two already related, and one that should appeal to those who enjoy combinatorial puzzles. This is what happened:

The inspector decided to stop off in Paris to attend to a few matters, and when he had finished he took a train from Paris to Calais, planning to cross the Channel to Dover. But, just as he got off at Calais, he was met by a French police officer who handed him a wire from Monte Carlo, begging him to come at once to help solve an "important problem."

"Oh, heavens," thought Craig, "I'll never get home at this rate!"

Still, duty was duty, and so Craig completely changed his plans, went to Monte Carlo, and was met at the station by an official named Martinez, who promptly took him to one of the banks.

"The problem is this," explained Martinez. "We have lost the combination to our biggest safe and to blow it open would be prohibitively expensive!"

"How ever did *that* happen?" asked Craig.

"The combination is written on only one card, and one of the employees carelessly left it inside the safe when he locked it!"

"Good heavens!" exclaimed Craig. "No one remembers the combination?"

"Absolutely no one," sighed Martinez. "And the worst of it is that if the wrong combination is used, the lock will be jammed permanently. Then there will be no recourse left but to blow open the safe, which, as I said, just isn't feasible—not only because of the cost of the mechanism but also because some extremely valuable and highly fragile materials are stored in it."

"Now, just a minute!" said Craig, "how can it be that you use a lock mechanism that can be permanently damaged by a wrong combination?"

"I was very much against purchasing the lock," said Martinez, "but I was overruled by the board of directors. They claimed that the mechanism had some uniquely valuable features which more than compensated for the disadvantage of possibly ruining it by using the wrong combination."

"This is really the most ridiculous situation I've ever heard!" said Craig.

"I heartily agree!" cried Martinez. "But what is to be *done?*"

"Frankly, I can't think of anything," replied Craig, "and *I* certainly cannot be of any help, since there are no clues. I'm very much afraid I have made this trip for nothing!"

"Ah, but there are clues!" said Martinez, a little more brightly. "Otherwise I would never have put you to the trouble of coming here."

"Oh?" said Craig.

"Yes," said Martinez. "Some time ago we had a very interesting though rather queer employee, a mathematician particularly interested in combinatorial puzzles. He took a keen interest in combination locks and studied the mechanism of this safe with great care. He pronounced it the most unusual and clever locking mechanism he had ever seen. He was constantly inventing puzzles, with which he amused many of us, and once he wrote a paper listing several properties of the locking mechanism, and asserting that from these properties we could actually *deduce* a combination that would open the safe. He gave this to us as a recreational puzzle, but it was far too difficult for any of us to solve, so we soon forgot it."

"And where is this paper?" asked Craig. "I suppose it is also locked up in the safe with the card bearing the combination?"

"Happily, no," said Martinez, as he produced the manuscript from his desk drawer. "Fortunately, I kept it in here."

Inspector Craig studied the manuscript carefully.

"I can see why none of you solved the puzzle; it appears extremely difficult! Wouldn't it be easier simply to contact the author? Surely he remembers or could reconstruct the combination, couldn't he?"

"He worked here under the name of 'Martin Farkus,' but that was probably an assumed name," replied Martinez. "No efforts to find him have been successful."

"Hm!" replied Craig, "I guess the only alternative is to try and solve this puzzle, but it might take weeks or several months."

"There is one more thing I must tell you," said Martinez. "It is absolutely imperative that the safe be opened by June first; it contains some state documents that *have* to be produced on the morning of June second. If we cannot find the combination by then, we will be forced to blow open the safe regardless of cost. The document itself won't be hurt by the explosion, since it is in a very stout inner safe, as far as possible from the door of the outer safe. And as for the other items—well, this document comes first! But it would be worth quite a sum of money to us not to have to resort to that alternative!"

"I'll see what I can do," said Craig, rising. "I can't promise you anything, though of course I'll do my best."

Now, let me tell you about the contents of Farkus's manuscript. To begin with, the combinations used letters, not numbers. And so by a *combination,* we will mean any string of any of the twenty-six capital letters of the alphabet. It can be of any length and contain any number of letters occurring any number of times; for example, BABXL is a combination; so is XEGGEXY. Also, a letter standing alone counts as a combination (a combination of length 1). Now, certain combinations will open the lock, certain ones will jam the lock, and the remaining combinations have no effect on the mechanism whatever. Those that have no effect on the mechanism are called *neutral.* We shall use the small letters x and y to represent arbitrary combinations, and by xy is meant combination x followed by combination y; for example, if x is the combination GAQ and y is the combination DZBF, then xy is the combination GAQDZBF. By the *reverse* of a combination is meant the combination written backwards; for example, the reverse of the combination BQFR is RFQB. By the *repeat,* xx, of a combination x is meant the combination followed by itself; for example, the repeat of BQFR is BQFRBQFR.

Now, Farkus (or whatever his real name was) referred to certain combinations as being *specially related* to others (or possibly to themselves), but he never defined what he meant by this term. Nevertheless, he listed enough properties of this "special relation" (whatever that might be) to enable a clever person to find a combination that opens the lock! He listed the following five key properties (which he said held for any combinations *x* and *y*):

Property Q: For any combination *x*, the combination Q*x*Q is specially related to *x*. (For example, QCFRQ is specially related to CFR.)

Property L: If *x* is specially related to *y*, then L*x* is specially related to Q*y*. (For example, since QCFRQ is specially related to CFR, then LQCFRQ is specially related to QCFR.)

Property V (the reversal property): If *x* is specially related to *y*, then V*x* is specially related to the reverse of *y*. (For example, since QCFRQ is specially related to CFR, then VQCFRQ is specially related to RFC.)

Property R (the repetition property): If *x* is specially related to *y*, then R*x* is specially related to *yy* (the repeat of *y*). (For example, since QCFRQ is specially related to CFR, then RQCFRQ is specially related to CFRCFR. Also—as we saw in the example accompanying Property V—VQCFRQ is specially related to RFC, and hence RVQCFRQ is specially related to RFCRFC.)

Property Sp: If *x* is specially related to *y*, then if *x* jams the lock, *y* is neutral, and if *x* is neutral, then *y* jams the lock. (For example, we have seen that RVQCFRQ is specially related to RFCRFC. Therefore, if RVQCFRQ should jam the lock, then RFCRFC would have no effect on the mechanism, and if LVQCRFQ has no effect on the mechanism, then RFCRFC jams the lock.)

From these five conditions, it is indeed possible to find a combination that opens the lock. (The shortest one I know is of length 10, and there are others.)

Now, the reader is hardly expected to solve this puzzle at this point; there is a whole theory behind this mechanism which will gradually unfold in the course of the next few chapters. This theory is related to some very interesting discoveries in mathematics and logic that will be apparent later on.

As a matter of fact, Craig worked on this puzzle for several days after his interview with Martinez, but was unable to solve it.

"No sense remaining here any longer," thought Craig. "I have no idea how long this will take, and I might just as well work on it at home."

And so Craig went back to London. That the puzzle ever did get solved was due not only to the ingenuity of Craig and two of his friends (whom we shall meet presently), but also to the remarkable concatenation of circumstances about to unfold.

◆ 9 ◆

A Curious Number Machine

After Craig's return to London, he at first spent a good deal of time on the Monte Carlo lock puzzle. Then, since he was getting nowhere, he decided that it might be best to rest a while from the problem and went to visit an old friend named Norman McCulloch whom he had not seen for years. He and McCulloch had been fellow students at Oxford, and Craig recalled him in those days as a delightful, if somewhat eccentric, chap who was constantly inventing all sorts of curious gadgets. Now, this whole story takes place in the days before modern computers were invented, but McCulloch had put together a crude mechanical computer of a sort.

"I've been having ever so much fun with this device," explained McCulloch. "I've not yet found any practical use for it, but it has some intriguing features."

"What does it do?" asked Craig.

"Well," replied McCulloch, "you put a number into the machine, and after a while a number comes out of the machine."

"The same number or a different one?" asked Craig.

"That depends on what number you put in."

"I see," replied Craig.

"Now," continued McCulloch, "the machine doesn't accept *all* numbers—only some. Those which the machine accepts, I call *acceptable* numbers."

"That sounds like perfectly logical terminology," said Craig, "but I would like to know which numbers are acceptable and which are not. Is there a definite law concerning this? Also, is there a definite law concerning what number comes out once you have decided what acceptable number to put in?"

"No," replied McCulloch, "*deciding* to put the number in is not enough; you have actually to put the number in."

"Oh, well, of course!" said Craig. "I meant to ask whether once the number has been put in, if it is definitely determined what number comes out."

"Of course it is," replied McCulloch. "My machine is not a random device! It operates according to strictly deterministic laws.

"Let me explain the rules," he continued. "To begin with, by a *number* I mean a positive whole number; my present machine doesn't handle negative numbers or fractions. A number N is written in the usual way as a string of the digits 0,1,2,3,4,5,6,7,8,9. However, the only numbers my machine handles are those in which 0 does not occur; for example, numbers like 23 or 5492, but not numbers like 502 or 3250607. Given two numbers *N,M*—now by NM I *don't* mean N times M! By NM I mean the number obtained by first writing the digits of N in the order in which they occur, and then following it by the digits of M; so, for example, if N is the number 53 and M is the number 728, by NM I mean the number 53728. Or if N is 4 and M is 39, by NM I mean 439."

"What a curious operation on numbers!" exclaimed Craig in surprise.

"I know," replied McCulloch, "but this is the operation the machine understands best. Anyway, let me explain to you the rules of operation. I say that a number X *produces* a number Y, meaning that X is acceptable and that when X is put into the machine, Y is the number that comes out. The first rule is as follows:

"***Rule 1:*** For any number X, the number 2X (that is, 2 *followed* by X, not 2 times X!) is acceptable, and 2X produces X.

"For example, 253 produces 53; 27482 produces 7482; 23985 produces 3985, and so forth. In other words, if I put a number 2X into the machine, the machine erases the 2 at the beginning and what is left—the X—comes out."

"That's easy enough to understand," replied Craig. "What are the other rules?"

"There is only one more rule," replied McCulloch, "but first let me tell you this: For any number X, the number X2X plays a particularly prominent role; I call the number X2X the *associate* of the number X. So, for example, the associate of 7 is 727; the associate of 594 is 5942594. Now, here is the other rule:

"***Rule 2:*** For any numbers X and Y, if X produces Y, then 3X produces the associate of Y.

"For example, 27 produces 7, by Rule 1; therefore 327 produces the associate of 7, which is 727. Thus 327 produces 727. Again, 2586 produces 586; hence 32586 produces the associate of 586, which is 5862586."

At this point, McCulloch fed the number 32586 into the machine and, after much groaning and squeaking, the number 5862586 finally did come out.

"Machine needs a little oiling," commented McCulloch. "But let's consider another example or two to see if you have fully grasped the rules. Suppose I put in 3327; what will come out? We already know that 327 produces 727; so 3327 produces the associate of 727, which is 7272727. What number does 33327 produce? Well, since 3327 produces 7272727 (as we have just seen), then 33327 must produce the associate of 7272727, which is 727272727272727. As another example, 259 produces 59; 3259 produces 59259; 33259 produces 59259259259; 333259 produces 5925925925925925925925 9."

"I see," said Craig, "but the only numbers you have mentioned so far which seem to 'produce' anything are numbers beginning with either 2 or 3. What about numbers beginning, say, with 4?"

"Oh, the only numbers accepted by this machine are those beginning with 2 or 3, and not even all of those are acceptable. I am planning one day to build a larger machine which accepts more numbers."

"What numbers beginning with 2 or 3 are not acceptable?" asked Craig.

"Well, 2 alone is not acceptable, since it does not come within the scope of either Rule 1 or Rule 2, but any multidigital number beginning with 2 is acceptable. No number consisting entirely of 3's is acceptable. Also 32 is not acceptable, nor is 332, nor any string of 3's followed by 2. But for any number X, 2X is acceptable; 32X is acceptable; 332X and 3332X are acceptable; and so forth. In short, the only acceptable numbers are 2X, 32X, 332X, 3332X, and any string of 3's followed by 2X. And 2X produces X; 32X produces the associate of X; 332X produces the associate of the associate of X—which it is convenient to call the *double associate* of X; 3332X produces the associate of the associate of the associate of X—this number I call the *triple associate* of X—and so on."

"I fully understand," said Craig, "and now I would like to know just what are the curious features of this machine to which you have alluded?"

"Oh," replied McCulloch, "it leads to all sorts of curious combinatorial puzzles—here, let me show you some!"

1

"To begin with a simple example," said McCulloch, "there is a number N which produces itself; when you put N into the machine, out comes the very same number N. Can you find such a number?"

2

"Very good," said McCulloch, after Craig showed him his solution. "And now for another interesting feature of this machine: There is a number N which produces its own associate—in other words, if you put N into the machine, the number N2N comes out. Can you find such a number?"

Craig found this puzzle more difficult, but he managed to solve it. Can you?

3

"Excellent!" said McCulloch. "But there is one thing I would like to know: how did you go about finding this number? Was it just trial and error, or did you have some systematic plan? Also, is the number you found the only number that produces its own associate, or are there others?"

Craig then explained his method for finding the number N in the last problem, and also answered McCulloch's question as to whether there were other possible solutions. The reader should find Craig's analysis here to be of considerable interest, and it facilitates, moreover, the solutions of several other puzzles of the present chapter.

4

"Apropos of my last question," said McCulloch, "how did you solve the first problem? Is there more than one number that produces itself?"

Craig's answer is given in the solutions.

5

"Next," said McCulloch, "there is a number N which produces 7N (that is, 7 followed by N). Can you find it?"

6

"Now, let's consider another question," said McCulloch. "Is there a number N such that 3N produces the associate of N?"

7

"And is there an N," asked McCulloch, "which produces the associate of 3N?"

8

"A particularly interesting feature of this machine," said McCulloch, "is that for any number A there is some number

Y which produces AY. How do you prove this, and, given a number A, how do you find such a number Y?"

Note: This principle, simple as it is, is more important than McCulloch realized at the time! It will crop up several times in the course of this book. We shall call it *McCulloch's Law.*

9

"Now," continued McCulloch, "given a number A, is there necessarily some Y that produces the associate of AY? For example, is there a number that produces the associate of 56Y, and if so, what number does this?"

10

"Another interesting thing," said McCulloch, "is that there is a number N that produces its own double associate. Can you find it?"

11

"Also," said McCulloch, "given any number A there is a number X that produces the double associate of AX. Can you see how to find such an X, given the number A? For example, can you find an X that produces the double associate of 78X?"

Here are some more problems that McCulloch gave Craig on this day. (Except for the last of them, they are not of theoretical importance, but the reader might have fun playing with them.)

12

Find a number N such that 3N produces 3N.

13

Find a number N such that 3N produces 2N.

14

Find a number N such that 3N produces 32N.

15

Is there an N such that NNN2 and 3N2 produce the same number?

16

Is there an N whose associate produces NN? Is there more than one such N?

17

Is there an N such that NN produces the associate of N?

18

Find an N such that the associate of N produces the double associate of N.

19

Find an N that produces N23.

20 ◆ A Negative Result

"You know," said McCulloch, "for quite some time I have been trying to find a number N that produces the number N2, but so far all my attempts have failed. I wonder whether in fact there is no such number or whether I just haven't been clever enough to find one!"

This problem immediately engaged Craig's attention. He took out a notebook and pencil and started working on it. After a while he said, "Don't lose any more time looking for such a number: it cannot possibly exist!"

How did Craig know this?

◆ SOLUTIONS ◆

1 ◆ One such number is 323. Since 23 produces 3 (by Rule 1), then, by Rule 2, 323 must produce the associate of 3, which is 323—the very same number!

Are there other such numbers? For Craig's answer, see the solution to Problem 4.

2 ◆ The number Craig found was 33233. Now, any number of the form 332X produces the double associate of X, so 33233 produces the double associate of 33—that is, the associate of the associate of 33. Now, the associate of 33 is the original number 33233; hence the double associate of 33 is

the associate of 33233. Thus 33233 produces the associate of 33233—that is, it produces its own associate.

How was this number found, and is it the only solution? Craig gives his answers to these questions in the solution to the next problem.

3 • Here is how Craig found a solution to Problem 2 and also settled the question of whether or not there are any other solutions. I shall give his explanation in his own words:

"My problem was to find a number N that produces N2N. This N must be one of the forms 2X, 32X, 332X, 3332X, etc., and I must discover X. Could a number of the form 2X work? Clearly not, since 2X produces X, which is obviously shorter (has fewer digits) than the associate of 2X. So no number of the form 2X could possibly work.

"What about a number of the form 32X? It also produces a number which is too short; it produces the associate of X, which is obviously shorter than the associate of 32X.

"What about a number of the form 332X? Well, it produces the double associate of X, which is X2X2X2X, whereas what is required is to produce the associate of 332X, which is 332X2332X. Now, can X2X2X2X be the same number as 332X2332X? What about the comparative lengths? Well, letting h be the number of digits in X, the number X2X2X2X has $4h + 3$ digits (since there are four X's and three 2's), whereas 332X2332X has $2h + 7$ digits. Can $4h + 3 = 2h + 7$? Yes, if $h = 2$, but for no other h. So lengthwise, a number of the form 332X may be a possibility, but only if h has two digits.

"Are there any other possibilities? What about a number of the form 3332X? It produces the triple associate of X, which is X2X2X2X2X2X2X2X, whereas what is required is to produce the associate of 3332X, which is 3332X23332X. Could these numbers be the same? Again, letting h be the length of X, the number X2X2X2X2X2X2X2X has $8h + 7$ digits, whereas 3332X23332X has $2h + 9$ digits. The only solution to the equation $8h + 7 = 2h + 9$ is that $h = ⅓$, so there is no whole number h that will make $8h + 7 = 7h + 9$; therefore no number of the form 3332X can work.

"What about a number of the form 33332X? It produces the quadruple associate of X, which has a length of $16h + 15$, whereas the associate of X has a length of $2h + 11$. Of course, for any positive integer h, $16h + 15$ is larger than $2h + 11$, so a number of the form 33332X produces something too large.

"If we take a number beginning with five 3's instead of four, the disparity between the lengths of the number it is supposed to produce and the number it actually produces is even greater, and if we take a number beginning with six or more 3's, the disparity is greater yet. Therefore, we are back to 332X as the only possible solution to the problem, so X must be a two-digit number. Thus, the desired N must be of the form 332*ab*, where *a* and *b* are single digits to be determined. Now, 332*ab* produces the double associate of *ab*, which is *ab*2*ab*2*ab*2*ab*. It is *desired* that 332*ab* produce the associate of 332*ab*, which is 332*ab*2332*ab*. Can these two numbers be the same? Let us compare them digit by digit:

*ab*2*ab*2*ab*2*ab*
332*ab*2332*ab*

"Comparing the first digits of each number, we see that *a* must be 3. Comparing the second digits, *b* must also be 3. So N = 33233 is a solution, and is the only possible solution."

4 ◆ "To tell you the truth," said Craig, "I solved the first problem by intuition; I didn't find the number 323 by any

systematic method. Also, I have not yet considered whether there is any other number that produces itself.

"But I don't think this should be too difficult to settle: Let's see now, could a number of the form 332X work? It would produce the double associate of X, which is X2X2X2X, which has a length of $4h + 3$, with h being the length of X. But what is required is to produce 332X, which has a length of $h + 3$. Obviously, $4h + 3$ is greater than $h + 3$, if h is a positive number, so 332X produces a number that is too large. What about 3332X, or some number beginning with four or more 3's? No, the disparity would be greater yet; the only possibility is a number of the form 32X (a number of the form 2X is clearly no good; it can't produce 2X, since it produces X). Now, 32X produces X2X, and what is required is that it produce itself, which is 32X. So 32X must be the same as X2X. Letting h be the length of X, 32X has a length of $h + 2$, and X2X has a length of $2h + 1$. So $2h + 1 = h + 2$; this means that h must be 1. So X is a single digit. Now, for what digit a is it the case that $a2a = 32a$? Obviously, a must be 3. Hence 323 is the only solution."

5 • Take N to be 3273. It produces the associate of 73, which is 73273, which is 7N. So 73273 is a solution. (It is, in fact, the only solution, as can be shown by a comparative-lengths argument of the type considered in the last two problems.)

6 • Since 323 produces itself, then 3323 must produce the associate of 323. So, letting N = 323, 3N produces the associate of N. (It is the only solution.)

7 • The solution is 332333. Let us check: Let N be the number 332333. It produces the double associate of 333, which is the associate of 3332333—in other words, the associate of 3N.

8 ◆ This obviously is a straightforward generalization of Problem 5: We saw that for N = 3273, N produces 7N. There is nothing special about 7 that makes this work; for any number A, if we let Y = 32A3, Y produces AY (because it produces the associate of A3, which is A32A3, which is AY). So, for example, if we want a number Y that produces 837Y, we take Y to be 328373.

This fact will subsequently turn out to be of considerable theoretical importance!

9 ◆ The answer is *yes;* take Y to be 332A33. It produces the double associate of A33, which is the associate of A332A33. But A332A33 is AY, so Y produces the associate of AY.

For the particular example suggested by McCulloch—to find a number Y that produces the associate of 56Y—the solution is Y = 3325633.

10 ◆ The solution is 3332333. It produces the triple associate of 333, which is the double associate of the associate of 333. Now, the associate of 333 is 3332333, so 3332333 produces the double associate of 3332333.

The following general pattern should be noted: 323 produces itself; 33233 produces its own associate; 3332333 produces its own double associate. Also, 333323333 produces its own triple associate, 33333233333 produces its own quadruple associate, and so forth (as the reader can check for himself).

11 ◆ The solution is X = 3332A333. It produces the triple associate of A333, which is the double associate of the associate of A333. Now, the associate of A333 is A3332A333, which is AX. So X produces the double associate of AX.

For the particular example, A is 78, so the solution is 333278333.

12 • Obviously, the answer is 23. (We already know that 323 produces 323, so, letting N = 23, 3N produces 3N.)

13 • The answer is 22.

14 • The answer is 232.

15 • Of course: N = 2.

16 • Any string of 2's will work.

17 • Yes, 32 works.

18 • Take N = 33.

19 • Take N = 32323.

20 • As the reader can verify for himself, any number N beginning with two or more 3's will produce a number of greater length than that of N2 (for example, if N is of the form 332X, and h is the length of X, N produces the double associate of X, which has a length of $4h + 3$, whereas N2 has a length of $h + 4$). Also, no N of the form 2X could work, so if there is any N that produces N2, it must be of the form 32X. Now, 32X produces X2X, and what is required is to produce 32X2. If X2X is the same number as 32X2, then, letting h be the length of X, it must be that $2h + 1 = h + 3$, which means $h = 2$. So the only number that could work (if there is one) must be of the form 32*ab*, where a and b are single digits to

be determined. Now, 32*ab* produces *ab*2*ab,* and what is required is to produce 32*ab*2. So, can *ab*2*ab* be the same number as 32*ab*2? Let us compare them digit by digit:

*ab*2*ab*
32*ab*2

Comparing the first digits, we get $a = 3$; comparing the third digits, we find $a = 2$—and so the problem is impossible. There is no N that produces N2!

◆ 10 ◆
Craig's Law

A couple of weeks later, Craig paid another visit to McCulloch.

"I heard you have enlarged your machine," said Craig, "and some mutual friends have told me that your new machine does some very interesting things. Is that true?"

"Ah, yes!" replied McCulloch, with an air of pride. "My new machine obeys Rules 1 and 2 of my old machine, and in addition, it has two other rules. But I've just brewed some tea—let's have some before I show you the new rules."

After an excellent tea, complete with delicious hot-buttered crumpets, McCulloch began:

"By the *reverse* of a number, I mean the number written backwards; for example, the reverse of 5934 is 4395. Now, here is the first of the additional rules:

"*Rule 3:* For any numbers X and Y, if X produces Y, then 4X produces the reverse of Y.

"Let me illustrate," said McCulloch. "Pick a number Y at random."

"All right," said Craig, "suppose we take 7695."

"Very good," said McCulloch. "Let's take an X which produces 7695—we'll take 27695—and put 427695 into the machine and see what happens."

McCulloch then put 427695 into the machine and, sure enough, out came 5967—the reverse of 7695.

"Before I show you the next rule," said McCulloch, "let me show you some of the things that can be done with this rule—together, of course, with Rules 1 and 2."

1

"You recall," said McCulloch, "that the number 323 produces itself. Also, with my old machine—which didn't have Rule 3 built into it, only Rules 1 and 2—the number 323 was the *only* number that produced itself. With my present machine, the situation is different. Can you find another number that produces itself? Also, how many such numbers are there?"

It didn't take Craig too long to solve this. Can you do it? (The answer, in Craig's own words, is given in the solutions.)

2

"That was excellent," said McCulloch, after Craig had completed his exposition. "Let me give you another problem: I call a number *symmetric* if it reads the same both forwards and backwards—that is, if it is equal to its own reverse. Numbers like 58385 or 7447, for example, are symmetric. Numbers that are not symmetric I call *nonsymmetric*—numbers like 46733 or 3251. Now, there obviously is a number that produces its own reverse—namely, 323—because 323 both produces itself and is symmetric. With my first machine, which did not have Rule 3, there was no nonsymmetric number that produced its own reverse. But with Rule 3, there *is* one—in fact, several. Can you find one?"

3

"And then," said McCulloch, "there are numbers that produce the associates of their own reverse. Can you find one?"

"And now," said McCulloch, "here is the second new rule:

"*Rule 4:* If X produces Y, then 5X produces YY.

"I refer to YY as the *repeat* of Y."

McCulloch then gave Craig the following problems.

4

Find a number that produces its own repeat.

5

Find a number that produces the reverse of its own repeat.

6

"That's curious," said McCulloch after Craig had solved Problem 5. "I obtained a different solution—also one with seven digits."

There are indeed two seven-digit numbers each of which produces the reverse of its own repeat. Can you find the second of these?

7

"For any number X," said McCulloch, "52X obviously produces the repeat of X. Can you find a number X such that 5X produces the repeat of X?"

Craig thought about this for a bit and suddenly burst out laughing; the solution was so obvious!

8

"And now," said McCulloch, "there is a number that produces the repeat of its associate. Can you find it?"

9

"Also," said McCulloch, "there is a number that produces the associate of its own repeat. Can you find it?"

OPERATION NUMBERS

"You know," said Craig quite suddenly, "I just realized that almost all these problems can be solved by one general principle! Your machine has a very pretty property; once this is realized, it is possible to solve not only the problems you have given me but an infinite host of others!

"For example," continued Craig, "there must be a number that produces the repeat of the reverse of its own associate, and a number that produces the associate of the repeat of its own reverse, and a number that—"

"How extraordinary!" interrupted McCulloch. "I was looking for such numbers but couldn't find them. What are they?"

"You'll be able to find them within seconds, once I tell you this law!"

"What *is* the law?" pleaded McCulloch.

"Indeed," continued Craig, who was greatly enjoying McCulloch's mystification, "I can even give you a number X that produces the repeat of the reverse of the double associate of X, or a number Y that produces the reverse of the double associate of YYYY, or a number Z that—"

"Enough!" cried McCulloch. "Why don't you just tell me what this law is and leave the applications till later?"

"Fair enough!" replied Craig.

At this point, the inspector picked up a pad of paper that was lying on the table, took out a pencil, and sat McCulloch down beside him so that his friend could see what he was writing.

"To begin with," said Craig, "I presume that you are familiar with the notion of an *operation* on numbers; for example, the operation of adding 1 to a number, or multiplying a number by 3, or squaring a number; or, what is more relevant for your machine, taking the *reverse* of a number, or the *repeat* of a number, or the *associate* of a number, or perhaps a more complex operation like taking the reverse of the repeat of the associate of a number. Now, I shall use the letter F to stand for some given arbitrary operation, and for any number X, by F(X)—read 'F of X'—I mean the result of performing the operation F on the number X. This, as of course you know, is standard mathematical practice. So, for example, if F is the reverse operation, F(X) is the reverse of X; if F is the repetition operation, F(X) is the repeat of X, and so forth.

"Now, there are certain numbers—any number, in fact, composed of the digits 3, 4, or 5—which I shall call *operation numbers*, since they determine operations that your machine can perform: Let M be any number composed of the digits 3, 4, or 5, and let F be any operation. I will say that M *determines* the operation F, meaning that for any two numbers X and Y where X produces Y, the number MX must produce F(Y). For example, if X produces Y, then 4X produces the reverse of Y—by Rule 3—and so I say that the number 4 *determines* or *represents* the *reversal* operation; similarly, by Rule 4, the number 5 determines the *repetition* operation. The number 3 determines the *association* operation, that is, the operation of taking the associate of a number. Now, suppose F is the operation which, when applied to any number X, yields the associate of the repeat of X. In other words, F(X) is the associate of the repeat of X. Is there a number M that represents this operation, and if so, what is that number?"

"Obviously 35," answered McCulloch, "because if X produces Y, 5X produces the repeat of Y; hence 35X produces the associate of the repeat of Y. Thus, 35 represents the operation of taking the associate of the repeat of a number."

"Right!" replied Craig. "I have now defined what it means for an operation number M to represent an operation, and this operation I will refer to as *Operation M*. So, for example, Operation 4 is the reversal operation; Operation 5 is the repetition operation; Operation 35 is the operation of taking the associate of the repeat, and so forth. . . .

"Here is a question," he continued. "Is it possible for two different numbers to represent the same operation? That is, can there be operation numbers M and N such that M is different from N, yet operation M is the same as operation N?"

McCulloch thought for a moment. "Oh, of course!" he said. "The numbers 45 and 54 are different, but they determine the same operation, since the reverse of the repeat of a number is the same as the repeat of its reverse."

"Good," replied Craig, "though I was thinking of a different example: To begin with, what operation does 44 represent?"

"Well," said McCulloch, "Operation 44 applied to X gives the reverse of the reverse of X, which is X itself. I don't know what name to give to an operation which, when applied to any number X, simply gives X itself."

"In mathematics, it is commonly called the *identity* operation," remarked Craig, "and so the number 44 determines the identity operation. But so does 4444, or any number composed of an even number of 4's, therefore there are infinitely many different numbers that represent the identity operation. And, more generally, given any operation number M, then M followed or preceded by an even number of 4's (or both) represents the same operation as does M alone."

"I see that," said McCulloch.

"And now," said Craig, "given an operation number M and any number X, I want a convenient notation for the result of applying operation M to the number X; I shall simply write this as 'M(X).' For example, 3(X) is the associate of X; 4(X) is the reverse of X; 5(X) is the repeat of X; 435(X) is the reverse of the associate of the repeat of X. Is this notation clear?"

"Oh, yes," replied McCulloch.

"You won't, I trust, ever confuse the notation M(X) with MX; the former means the result of applying operation M to X; the latter is the number M *followed by* the number X, and these are very different things! For example, 3(5) is not 35, but 525."

"I understand that," said McCulloch, "but can it ever happen—by some sort of coincidence—that M(X) is the same as MX?"

"Good question," replied Craig. "I'll have to think about that!"

"First, let's have another cup of tea," suggested McCulloch.

"Excellent!" replied Craig.

While our two friends are enjoying their tea, I'd like to give you some puzzles about operation numbers; they will afford good practice in the use of the notation M(X), which will play a vital role later on.

10

The answer to McCulloch's last question is yes: There *are* an operation number M and a number X such that M(X) = MX. Can you find them?

11

Is there an operation number M whereby M(M) = M?

12

Find an operation number M and a number X whereby M(X) = XXX.

13

Find an operation number M and a number X whereby M(X) = M + 2.

14

Find M and X such that M(X) is the repeat of MX.

15

Find operation numbers M and N such that M(N) is the repeat of N(M).

16

Find two *distinct* operation numbers M and N such that M(N) = N(M).

17

Can you find two operation numbers M and N such that M(N) = N(M) + 39?

18

What about two operation numbers M and N such that M(N) = N(M) + 492?

19

Find two *distinct* operation numbers M and N such that M(N) = MM and N(M) = NN.

CRAIG'S LAW

"You still haven't told me the principle you claim you have discovered," said McCulloch, after they had finished their tea. "I presume that your talk of operation numbers and operations is leading up to this?"

"Oh, yes," replied Craig, "and I think you are now ready to grasp this law. Do you recall some of the earlier problems you gave me? For example, finding a number X that produced its own repeat. In other words, we wanted a number X that produces 5(X). Or, in finding a number X that produced its own associate, we wanted an X that produces 3(X). Or, again, a number X that produces the reverse of X is a number that produces 4(X). But all these are special cases of one general principle—namely, that for *any* operation number M, there must be an X that produces M(X)! In other words, given any operation F that your machine can perform—that is, any operation F which is determined by some operation number M—there must be an X that produces F(X).

"Moreover," continued Craig, "given an operation number M, we can find an X that produces M(X) by a very simple recipe. Once you know this general recipe, then, for example, you can find an X that produces 543(X), which solves the problem of finding an X that produces the repeat of the reverse of the associate of X, and also you can find an X that produces 354(X), which solves the other problem of finding a number that produces the associate of the repeat of its own reverse. Or, as I told you, I can find a number X that produces the repeat of the reverse of the double associate of X—in other words, an X which produces 5433(X). Without the recipe I have in mind, such problems can be exceedingly difficult, but with it they are child's play!"

"I am all ears," said McCulloch. "What is this remarkable recipe?"

"I am about to tell you," said Craig, "but first let's get one elementary fact absolutely straight—namely, that for any operation number M and for any numbers Y and Z, if Y produces Z, then MY produces M(Z). For example, if Y produces Z, then 3Y produces 3(Z)—the associate of Z; 4Y produces 4(Z); 5Y produces 5(Z); 34Y produces 34(Z). And likewise for *any* operation number M, if Y produces Z, MY produces M(Z). In particular, since 2Z is an example of some Y that produces Z, then it is always the case that M2Z produces M(Z). (For example, 32Z produces 3(Z)—the associate of Z; 42Z produces 4(Z)—and for any operation number M, M2Z produces M(Z).) Indeed, we could have defined M(Z) as the number produced by M2Z."

"I understand all that," said McCulloch.

"Well," replied Craig, "this fact is very easy to forget, so let me repeat it, and let us carefully make a note of it and remember it well!

"*Fact 1:* For any operation number M and any numbers Y and Z, if Y produces Z, then MY produces M(Z). (In particular, M2Z produces M(Z).)

"From this fact," Craig went on, "together with a fact which you discovered about your first machine and which also holds for your present machine, it easily follows that, given any operation number M, there must be some number X that produces M(X)—X produces the result of applying operation M to X. And, given M, such an X can be found by a simple general recipe."

20

Craig has discovered a basic principle that will henceforth be called *Craig's Law*—namely, that for any operation number M, there must be some number X that produces M(X). How do you prove Craig's Law and, given an M, how do you find such an X? For example, what X produces 543 (X)? Or, what X produces the repeat of the reverse of the associate of X? And what X produces the associate of the repeat of the reverse of X—that is, what X produces 354(X)?

"I have a few more problems I'd like you to see," said McCulloch, "but it's getting quite late. Why don't you stay the night? I can show you the problems tomorrow."

It happened that Craig was on vacation at the time, so he gladly accepted McCulloch's invitation.

SOME VARIANTS OF CRAIG'S LAW

Next morning, after a hearty breakfast (McCulloch was an excellent host!), McCulloch gave Craig the following problems:

21

Find a number X that produces 7X7X.

22

Find a number X that produces the reverse of 9X.

23

Find a number X that produces the associate of 89X.

"Very clever!" exclaimed Craig, after he had solved these problems. "None of these three problems can be solved by the law I gave you yesterday."

"That's right!" laughed McCulloch.

"And yet," said Craig, "all three can be solved by a common principle: In the first place, the particular numbers 7, 5, and 89 are quite arbitrary; given *any* number A, there is an X that produces the repeat of AX, and there is an X that produces the reverse of AX, and there is an X that produces the associate of AX. There is also an X that produces the repeat of the reverse of AX, or the reverse of the associate of AX—indeed, given *any* operation number M, and given any number A, there must be an X that produces M(AX)—that is, the number obtained by applying operation M to the number AX."

24

Craig, of course, was right: given any operation number M and any number A, there must be an X that produces M(AX). Let us call this principle *Craig's Second Law.* How do you prove this law, and, given an operation number M and a number A, how, explicitly, do you find an X that produces M(AX)?

25

"I just thought of another question," said McCulloch. "For any number X, let $\overleftarrow{X}$ represent the reverse of X. Can you find a number X that produces $\overleftarrow{X}67$? (That is, is there an X that produces the reverse of X followed by 67?) In general, is it true that for any number A there is an X that produces $\overleftarrow{X}A$?"

26

"Another question has occurred to me," said McCulloch. "Is there a number X that produces the *repeat* of $\overleftarrow{X}67$? More generally, is it true that for any A there is some X that produces the repeat of $\overleftarrow{X}A$? Still more generally, is it true that for any A and any operation number M there must be some X that produces $M(\overleftarrow{X}A)$?"

Discussion: Craig's Law holds not only for McCulloch's second machine but also for his first one—and, indeed, for any possible machine that obeys Rules 1 and 2. That is, however we extend McCulloch's first machine by adding new rules, the resulting mechanism is still subject to Craig's Law (in fact, to both of Craig's laws).

Craig's first law is related to a famous result in the theory of computable functions known as the *Recursion Theorem* (or sometimes as the *Fixed-Point Theorem*). McCulloch's Rules 1 and 2 are about the most economical ones I have ever seen achieve this result. They have another surprising property (related to another famous result in the theory of computable functions known as the *Double Recursion Theorem*), which will be explained in the next chapter. All this is relevant to the subjects of self-reproducing machines and cloning.

• SOLUTIONS •

1 • "With your present machine, there are *infinitely* many different numbers that will produce themselves," said Craig.

"Right!" said McCulloch. "How do you prove this?"

"Well," replied Craig, "let me call a number S an A-number if it has the property that for any numbers X and Y where X produces Y, SX produces the associate of Y. Now, before you added this new rule, 3 was the only A-number. But with your present machine, there are infinitely many A-numbers, and for *any* A-number S, the number S2S *must* produce itself, since S2S produces the associate of S, which is S2S."

"How do you know there is an infinite amount of A-numbers?" asked McCulloch.

"To begin with," replied Craig, "do you grant that for any numbers X and Y, if X produces Y then 44X will also produce Y?"

"Clever observation!" replied McCulloch. "Of course you are right: if X produces Y, then 4X produces the reverse of Y; hence 44X must produce the reverse of the reverse of Y, which is Y itself."

"Good," said Craig, "and so if X produces Y, 44X will also produce Y, and hence 344X will produce the associate of Y. Therefore, 344 is also an A-number. And since 344 is an A-number, then 3442344 must produce itself!"

"Very good!" said McCulloch. "So now we have two numbers—323 and 3442344—which produce themselves. How does this give us an infinite supply of such numbers?"

"Obviously," said Craig, "if S is an A-number, so is S44, because for any numbers X and Y, if X produces Y, then 44X also produces Y, and so S44X produces the associate of Y, since S is an A-number. So 3 is an A-number; hence so is 344; hence also 34444, and, in general, 3 followed by any *even* number of 4's is an A-number. So 323 produces itself; so does 3442344; so does 34444234444, and so forth. And thus we have infinitely many solutions.

"Incidentally," added Craig, "these are not the *only* solutions; the numbers 443 and 44443 are also A-numbers—indeed, any even number of 4's followed by 3 followed by an even number of 4's, such as 4434444 is an A-number, and so for every such number S, S2S produces itself."

2 ◆ 43243 is one solution: Since 243 produces 43, then 3243 produces the associate of 43. Therefore, 43243 must produce the reverse of the associate of 43—in other words, the reverse of 43243 (since 43243 is the associate of 43). So 43243 produces its own reverse.

At this point the reader may well be wondering how the number 43243 was found. Was it by a comparative-lengths argument? No, comparative-lengths arguments are quite unwieldy for proving things about this present machine. The solution was found by Craig's Law, as we shall see later in this chapter.

3 ◆ One solution is 3432343. We leave it to the reader to calculate the number produced by 3432343, and he will see that it is indeed the associate of the reverse of 3432343. (This solution was also found by using Craig's Law.)

4 ◆ 53253 works. (Craig's Law is again responsible for the answer.)

5 ◆ 4532453 is one solution.

6 ◆ 5432543 is another solution.

7 • Obvious, that is, once we know that some number produces itself. If X produces X, then of course 5X produces the repeat of X. So, for example, 5323 produces the repeat of 323.

8 • 5332533 is a solution. (Craig's Law again.)

9 • 3532353 is a solution; it was also found by Craig's Law. (I hope I am working up the reader's appetite to learn Craig's Law!)

10 • 5 (5) = 55. (Because 5 (5) is the repeat of 5.) So we take 5 for M and also 5 for X. (I never said that M and X must be different!)

11 • 4 (4) = 4. (Since 4 (4) is the reverse of 4, which is 4.) And so M = 4 is one solution. (Actually, any string of 4's would work.)

12 • Try M = 3, and X = 2. (3 (2) = 222.)

13 • 4 (6) = 6, and 6 = 4 + 2, so 4 (6) = 4 + 2. So M = 4 and X = 2.

14 • M = 55, X = 55 is a solution.

15 • M = 4, N = 44 is a solution.

16 • M = 5, N = 55 is a solution.

17 • M = 5, N = 4 is a solution.

18 • M = 3, N = 5 is a solution.

19 • M = 54, N = 45 is a solution.

20 ◆ Let M be any operation number. We know (Fact 1) that for any numbers Y and Z, if Y produces Z, then MY produces M(Z). Therefore (taking MY for Z), if Y produces MY, then MY must produce M(MY). Thus, taking X for MY, the number X will produce M(X)! So the problem boils down to finding some Y that produces MY. But this problem was solved in the last chapter (by McCulloch's Law)—namely, take 32M3 for Y! And so for X, we take M32M3, and X will produce M(X).

Let us double-check: Let X = M32M3. Since 2M3 produces M3, then 32M3 produces M32M3 (by Rule 2), and hence M32M3 produces M(M32M3); thus X produces M(X), where X is M32M3.

To consider some applications: To find an X that produces the repeat of X, we take 5 for M, and so the solution (or rather one solution) is 53253. To find an X that produces its own reverse, we take 4 for M, and X is then 43243. To find an X that produces the associate of the reverse of X, we take 34 for M, and one solution is 3432343.

For McCulloch's first problem—finding an X that produces the repeat of the reverse of X's associate—we take 543 for M (5 for repeat, 4 for reverse, and 3 for associate), and the solution is 543325433. (The reader can verify directly that 543325433 produces the repeat of the reverse of the associate of 543325433.) For McCulloch's second problem—finding an X that produces the associate of the repeat of the reverse of X—we take 354 for M and get the solution 354323543.

Craig's Law is really marvelous!

21, 22, 23, 24 ◆ Problems 21, 22, and 23 are all special cases of Problem 24; so let us first do Problem 24.

We are given an operation number M and an arbitrary number A, and we wish to find an X that produces M(AX).

The trick now is to find some Y that does not produce MY but that does produce AMY: Let's take 32AM3 for Y. Since Y produces AMY, then, by Fact 1, MY must produce M(AMY). Thus, with MY taken for X, X produces M(AX). Since we took 32AM3 for Y, our X is then M32AM3. And so M32AM3 is our desired solution.

To apply this to Problem 21, we first notice that 7X7X is simply the repeat of 7X, and so we want an X that produces the repeat of 7X—the repeat of AX, with A being 7. So A is 7, and we obviously take 5 for M (since 5 represents the repetition operation), and so the solution is 532753. (The reader can verify directly that 532753 does indeed produce the repeat of 7532753.) For Problem 22, A is 9, and we take 4 for M, and the solution is 432943. For Problem 23, A is 89, and we take 3 for M, so the solution is 3328933.

25 ◆ Yes, for any Number A there is an X that produces $\overleftarrow{X}$A—namely, 432$\overleftarrow{A}$43. (For this particular problem in which A is 67, $\overleftarrow{A}$ is 76, and so the solution is 4327643.)

26 ◆ For the most general case, the trick is to realize that $\overleftarrow{X}$A is the reverse of $\overleftarrow{A}$X, and so M($\overleftarrow{X}$A) = M4($\overleftarrow{A}$X). By Craig's Second Law, an X that produces M4($\overleftarrow{A}$X) is M432$\overleftarrow{A}$M43, so this is a solution. In particular, taking 5 for M and 67 for A, an X that produces the repeat of $\overleftarrow{X}$67 is 543276543 (as the reader can verify directly).

◆ 11 ◆
Fergusson's Laws

And now we come to a further interesting development concerning McCulloch's machines. About two weeks after the last episode, McCulloch received the following letter from Craig:

My dear McCulloch:

I am greatly intrigued by your number machines and so is my friend Malcolm Fergusson. Do you by any chance know Fergusson? He is actively engaged in research in pure logic and has himself constructed several logic machines. His interests, however, extend far and wide; for example, he is very interested in that variety of chess problems known as *retrograde analysis.* He also takes a keen interest in pure combinatory problems—the kind your machines so ably provide. I visited him last week and showed him all your problems, and he was most intrigued. I met him again three days later, and he made some remark to the effect that he suspected that both of your machines have some interesting additional properties which even the inventor did not realize! He was a bit vague about all this and said that he wanted more time to think the matter over.

Fergusson is coming to dine with me next Friday evening. Why don't you join us? I'm sure you will have much in common, and it might be very interesting to find out what he has in mind concerning your machines.

Hoping to see you then, I remain

Sincerely yours,

L. Craig

McCulloch promptly replied:

Dear Craig:

No, I have not met Malcolm Fergusson, but I have heard a good deal about him through mutual friends. Wasn't he a student of the eminent logician Gottlob Frege? I understand he is working on some ideas that are basic to the entire foundation of mathematics, and I certainly welcome this opportunity to meet him. Needless to say, I am very curious as to what he has in mind concerning my machines. I thank you for your invitation, which I happily accept.

Sincerely,

N. McCulloch

◊ ◊ ◊

The two guests arrived. After an excellent dinner (prepared by Craig's landlady, Mrs. Hoffman), the mathematical conversation began.

"I understand that you have constructed some logic machines," said McCulloch. "I would like to know more about them. Can you explain them to me?"

"Ah, that's a long story," replied Fergusson, "and I still haven't solved a basic question concerning their operation. Why don't you and Craig visit my workshop sometime? Then I can tell you the whole story. This evening, however, I would prefer to talk about *your* machines. As I told Craig a few days ago, they have certain properties of which I suspect that even you are unaware."

"What are these properties?" asked McCulloch.

1

"Well," replied Fergusson, "let's start with a concrete example employing your second machine: There are numbers X and Y such that X produces the reverse of Y and Y produces the repeat of X. Can you find them?"

Craig and McCulloch were enormously intrigued by this problem, and immediately set to work trying to solve it. Neither one succeeded. The problem is of course solvable, and the ambitious reader might care to try his hand at it. There is a basic underlying principle involved (which will be explained in this chapter), and once the reader knows it, he will be amused at how simple the matter really is.

2

"I am totally mystified," said Craig, after Fergusson showed them the solution. "I see that your solution works, but how did you ever find it? Did you just stumble on these numbers X and Y by accident, or did you have some rational plan for finding them? To me, it seems like a conjuring trick!"

"Yes," said McCulloch, "it's like pulling a rabbit out of a hat!"

"Ah, yes," laughed Fergusson, who was thoroughly enjoying the mystification, "only it seems I pulled *two* rabbits out of a hat, each of which had a curious effect on the other."

"That's certainly true!" replied Craig. "Only I would like to know how you knew which rabbits to pull!"

"Good question, good question!" replied Fergusson, more jubilant than ever. "Here—let's try another: Find two numbers X and Y such that X produces the repeat of Y and Y produces the reverse of the associate of X."

"Oh, no!" cried McCulloch.

"Just a moment," said Craig. "I think I'm beginning to get an idea: Do you mean to tell us, Fergusson, that for any two operations that the machine can perform—given any two operation numbers M and N—there must exist numbers X and Y such that X produces M(Y) and Y produces N(X)?"

"Precisely!" exclaimed Fergusson, "and so, for example, we can find numbers X and Y such that X produces the double associate of Y and Y produces the repeat of the reverse of X—or any other combination you can name."

"Now, that's remarkable!" cried McCulloch. "These last few days I have been trying to construct a machine with just this property; I had no idea that I already had one!"

"You most assuredly do," rejoined Fergusson.

"How do you prove this?" asked McCulloch.

"Well, let me build up to the proof gradually," replied Fergusson. "The heart of the matter really lies in your Rules 1 and 2. So let me first make some observations about your first machine—the one using just those rules. We'll start with a simple problem: Using just Rules 1 and 2, can you find two *distinct* numbers X and Y such that X produces Y and Y produces X?"

Both Craig and McCulloch promptly set to work on this problem.

"Oh, of course!" said Craig, with a chuckle. "It obviously follows from something McCulloch showed me some weeks ago."

Can you find such an X and Y?

3

"And now," said Fergusson, "for any number A, there are numbers X and Y such that X produces Y and Y produces AX. Given an A, can you see how to find such an X and Y? For example, can you find numbers X and Y such that X produces Y and Y produces 7X?"

"Are we still working with just Rules 1 and 2, or can we use Rules 3 and 4?" asked Craig.

"You only need Rules 1 and 2," replied Fergusson.

Craig and McCulloch went to work on the problem.

"I've got a solution!" said Craig.

4

"That's interesting," said McCulloch, after Craig showed his solution. "I found a different solution!"

There is indeed a second solution. Can you find it?

5

"And now," said Fergusson, "we come to a really vital property: From just Rules 1 and 2, it follows that for any two numbers A and B, there exist numbers X and Y such that X produces AY and Y produces BX. For example, there exist numbers X and Y such that X produces 7Y and Y produces 8X. Can you find them?"

6

"It easily follows," said Fergusson, "from the last problem—or perhaps even more simply from Craig's Second Law—that for any operation numbers M and N there must exist X and Y such that X produces M(Y) and Y produces N(X). This holds not only for your present machine but for *any* machine whose rules include at least Rules 1 and 2. With your present machine, for example, there are numbers X and Y such that X produces the reverse of Y and Y produces the associate of X. Can you find them?"

7

"That's extremely interesting," said McCulloch to Fergusson after he and Craig had solved the last problem, "and now the following question occurs to me: Does my machine obey a 'double' analogue of Craig's Second Law? That is, given two operation numbers M and N and two numbers A and B, do there necessarily exist numbers X and Y such that X produces M(AY) and Y produces N(BX)?"

"Oh, yes," replied Fergusson. "For example, there are numbers X and Y such that X produces the repeat of 7Y and Y produces the reverse of 89X."

Can you find such numbers?

8

"I've thought of another question," said Craig. "Given an operation number M and a number B, is there necessarily an X and a Y such that X produces M(Y) and Y produces BX? For example, are there numbers X and Y such that X produces the associate of Y and Y produces 78X?"

Are there?

9

"As a matter of fact," said Fergusson, "many other combinations are possible. Given any operation numbers M and N and any numbers A and B, you can find numbers X and Y that fulfil any of the following conditions:

(a) X produces M(AY) and Y produces N(X).
(b) X produces M(AY) and Y produces BX.
(c) X produces M(Y) and Y produces X.
(d) X produces M(AY) and Y produces X.

How do you prove these facts?"

10 • Triplicates and Beyond

"Well, I imagine we have combed through just about all the possibilities," said Craig.

"Not really," replied Fergusson. "What I have shown you so far is only the beginning. Did you know, for example, that there are numbers X, Y, and Z such that X produces the reverse of Y, Y produces the repeat of Z, and Z produces the associate of X?"

"Oh, no!" exclaimed McCulloch.

"Oh, yes," rejoined Fergusson. "Given any three operation numbers M, N, and P, there must be numbers X, Y, and Z such that X produces M(Y), Y produces N(Z), and Z produces P(X)."

Can the reader see how to prove this? In particular, what numbers X, Y, and Z are such that X produces the reverse of Y, Y produces the repeat of Z, and Z produces the associate of X?

"Of course," said Fergusson, after Craig and McCulloch had solved the problem. "All sorts of variants of this 'triple' law are possible. For example, given any three operation numbers M, N, and P and any three numbers A, B, and C, there are numbers X, Y, and Z such that X produces M(AY), Y produces N(BZ), and Z produces P(CX). This also holds if you leave out any one or two of the numbers A, B, and C. Also, we can find numbers X, Y, and Z such that X produces AY, Y produces M(Z), and Z produces N(BX)—all sorts of variants are possible. But these you can work out at your leisure.

"Also," he continued, "the same idea works with four or more operation numbers. For example, we could find numbers X, Y, Z, and W such that X produces 78Y, Y produces the repeat of Z, Z produces the reverse of W, and W produces the associate of 62X. The possibilities are really endless. It all stems from the surprising power inherent in Rules 1 and 2."

• SOLUTIONS •

1 • One solution is to take X = 4325243, Y = 524325243. Since 25243 produces 5243, then 325243 produces the associ-

ate of 5243, which is 524325243, which is Y. Since 325243 produces Y, then 4325243 produces the reverse of Y, but 4325243 is X. Thus X produces the reverse of Y. Also, Y obviously produces the repeat of X (because Y is 52X, and since 2X produces X, 52X produces the repeat of X). Thus X produces the reverse of Y and Y produces the repeat of X.

2 ◆ Craig recalled McCulloch's Law: that for any number A there is some number X (namely, 32A3) that produces AX. In particular, if we take 2 for A, there is a number X (namely, 3223) that produces 2X. And, of course, 2X in turn produces X. So 3223 and 23223 are one pair of numbers that works; 3223 produces 23223 and 23223 produces 3223.

3 ◆ Craig solved the problem in the following manner: He reasoned that all that was necessary was to find some X that produces 27X. Then, if we let Y = 27X, X produces Y and Y produces 7X. Also, he found that there *is* an X that produces 27X—namely, 32273. And so Craig's solution was X = 32273; Y = 2732273.

Of course, this works not only for the particular number 7 but for any number A: If we let X = 322A3 and Y = 2A322A3, X produces Y and Y produces AX.

4 ◆ McCulloch, on the other hand, went about the problem in the following way: He reasoned that all that was necessary was to find some Y that produces 72Y. Then, if we let X be 2Y, X produces Y and Y produces 7X. We know how to find such a Y: Take Y = 32723. So McCulloch's solution was X = 232723; Y = 32723.

5 ◆ All that is necessary is to find an X that produces A2BX. Then, if we let Y = 2BX, X produces AY and Y produces BX.

An X that produces A2BX is 32A2B3. And so a solution is X = 32A2B3, Y = 2B32A2B3. (For the special case A = 7, B = 8, the solution is X = 327283, Y = 28327283.)

6 ◆ Let us first solve this problem using Craig's Second Law, which, we recall, says that for any operation number M and any number A, there is a number X (namely, M32AM3) that produces M(AX). Now, take any two operation numbers M and N. Then by Craig's Second Law (taking N2 for A), there is a number X (namely, M32N2M3) that produces M(N2X). And, of course, N2X produces N(X). So, if we let Y be N2X, X produces M(Y) and Y produces N(X). Thus, a solution is X = M32N2M3 and Y = N2M32N2M3. (For the particular problem suggested by Fergusson, we take 4 for M and 3 for N, and the solution is X = 4323243, Y = 324323243. The reader can check directly that X produces the reverse of Y and Y produces the associate of X—the second half is particularly obvious).

We could have also gone about the problem in the following way: By the solution to Problem 5, we know that there are numbers Z and W such that Z produces NW and W produces MZ (namely, Z = 32N2M3, W = 2M32N2M3). Then, by Fact 1 of the last chapter, MZ produces M(NW) and NW produces N(MZ); so if we let X be MZ and Y be NW, X produces M(Y) and Y produces N(X). We thus get the solution X = M32N2M3, Y = N2M32N2M3.

7 ◆ We now need an X that produces M(AN2BX); such an X is M32AN2BM3, by Craig's Second Law. Then take N2BX for Y. Then X produces M(AY), and Y (which is N2BX) obviously produces N(BX). So the general solution (or at least one such) is X = M32AN2BM3, Y = N2BM32AN2BM3. For the specific problem given here, we obviously take 5 for M, 4 for N, 7 for A, 89 for B.

8 ◆ By Craig's Second Law, there is an X that produces M(2BX)—namely, X = M322BM3. Then let Y = 2BX. So X produces M(Y) and Y produces BX. For the specific problem here, we take 3 for M and 78 for B, getting the solution X = 33227833, Y = 27833227833.

9 ◆ (a) Take an X that produces M(AN2X) and take Y to be N2X. (We can take X to be M32AN23; Y = N2M32AN23.) Then X produces M(AY) and Y produces N(X).

(b) Now take an X that produces M(A2BX) and take Y to be 2BX. (So now a solution is X = M32A2B3; Y = 2BM32A2B3.)

(c) If X produces M(Y), and Y = 2X, we have a solution, so take X = M322M3; Y = 2M322M3.

(d) If X produces M(AY) and Y = 2X, we have a solution, so take X = M32A2M3 and Y = 2M32A2M3.

10 ◆ By Craig's Second Law, there is an X that produces M(N2P2X)—namely, X = M32N2P2M3. Let Y = N2P2X, so X produces M(Y). Let Z = P2X, so Y = N2Z; hence Y produces N(Z). And Z produces P(X).

And so the solution is explicitly X = M32N2P2M3, Y = N2P2M32N2P2M3, Z = P2M32N2P2M3.

For the particular problem, the solution is X = 432523243, Y = 52324232523243 and Z = 32432523243.

The reader can directly compute that X produces the reverse of Y, Y produces the repeat of Z, and Z produces the associate of X.

Incidentally, given any three numbers A, B, and C, we can find numbers U, V, and W such that U produces AV, V produces BW, and W produces CU: Just take a U that produces A2B2CU (if we use the recipe of Craig's Second Law, U = 32A2B2C3). Then let V = 2B2CU and W = 2CU. Then U produces AV, V produces BW, and W produces CU. If now

A, B, and C are operation numbers, take X = AV, Y = BW, and Z = CU, and X produces A(Y), Y produces B(Z), and Z produces C(X), and so we have an alternative method of solving the problem.

◆ 12 ◆

Interlude: Let's Generalize!

Two days after the last episode, Craig was suddenly and quite unexpectedly sent by Scotland Yard over to Norway on a case which, though interesting, need not concern us. While he is gone, I will take the opportunity to offer you something of my own thoughts about McCulloch's number machines. The reader who is very anxious to find the solution to the Monte Carlo lock puzzle may defer this chapter till later if he wishes.

Mathematicians are very fond of generalizing! It is typical for one mathematician X to prove a theorem, and six months after the theorem is published, for mathematician Y to come along and say to himself, "Aha, a very nice theorem X has proved, but *I* can prove something even more general!" So he publishes a paper titled "A Generalization of X's Theorem." Or Y might perhaps be a little more foxy and do the following: he first *privately* generalizes X's theorem, and then he obtains a special case of his own generalization, and this special case *appears* so different from X's original theorem that Y is able to publish it as a new theorem. Then, of course, another mathematician, Z, comes along who is haunted by the feeling that *somewhere* there lies *something* of an important nature common to both X's theorem and Y's theorem, and after much labor, he finds a common principle. Z then publishes a paper in which he states and proves this new general principle, and adds: "Both X's theorem and Y's theorem can be obtained as special cases of my theorem by the following arguments. . . ."

Well, I am no exception, and so I wish first to point out some features of McCulloch's machines that I doubt either McCulloch, Craig, or Fergusson realized, and then I would like to make a few generalizations.

The first thing that struck me when I reviewed the discussion of McCulloch's second machine was that once Rule 4 (the repetition rule) is introduced, we no longer need Rule 2 (the associate rule) to obtain laws like Craig's and Fergusson's! Indeed, consider a machine which uses only Rules 1 and 4: For such a machine we can find a number X that produces itself; we can find one that produces its own repeat; given any A, we can find a number X that produces AX; we can find an X that produces the repeat of AX or the repeat of the repeat of AX. Also, still supposing Rule 2 has been deleted from McCulloch's machine, we can find an X that produces its own reverse or an X that produces the repeat of its reverse or an X that produces the reverse of AX or an X that produces the repeat of the reverse of AX. Also, suppose we consider a machine that obeys McCulloch's Rules 1, 2, and 4 (leaving out Rule 3, the reversal rule). There are now two different ways to construct a number that produces its own associate; there are two ways to construct a number that produces its own repeat; two ways to construct a number that produces the associate of its repeat or the repeat of its associate.

Finally, given *any* machine satisfying at least Rule 1 and

Rule 4, Craig's laws and Fergusson's laws all hold. And so we could have given alternative solutions to most of the problems of the last two chapters, using Rule 4 instead of Rule 2. (Can the reader see how all this can be done? If not, it will all be explained below.)

I could say much more, but to make a long business short, I will summarize my main observations in the form of three facts:

Fact 1: Just as any machine obeying Rules 1 and 2 also obeys McCulloch's Law (that for any A there is some X which produces AX), so does any machine obeying Rules 1 and 4.

Fact 2: Any machine obeying McCulloch's Law also obeys Craig's two laws.

Fact 3: Any machine obeying *both* Craig's Second Law *and* Rule 1 must also obey all of Fergusson's laws.

Can the reader see how to prove all this?

◆ SOLUTIONS ◆

Let us first consider any machine obeying Rules 1 and 4: For any X, 52X produces XX; hence if we take 52 for X, we see that 5252 produces 5252. So we have a number that produces itself. Also, 552552 produces its own repeat. Also, for any A, to find an X that produces AX, take X to be 52A52 (it produces the repeat of A52, which is A52A52, which is AX). This proves Fact 1. (If we want an X that produces the repeat of AX, take 552A552 for X.)

Now, let us consider a machine obeying McCulloch's Rules 1, 3, and 4. A number that produces its own reverse is 452452 (it produces the reverse of the repeat of 452; in other words, the reverse of 452452). (Compare this with the former solution 43243.) A number that produces the repeat of its own reverse is 54525452. (Compare with the former solution 5432543.)

Now, consider a machine obeying Rules 1, 2, and 4. We know that 33233 produces its own associate, but so does 352352. As for an X that produces its own repeat, we already have the two solutions 35235 and 552552. As for an X that produces the associate of its repeat, one solution is 3532353; another is 35523552. As for a number that produces the repeat of its associate, one solution is 5332533, and another is 53525352.

Now, consider an arbitrary machine obeying at least Rule 1 and Rule 4 of McCulloch's machines. Given an operation number M, an X that produces M(X) is M52M52. (Compare this with the former solution M32M3, using Rule 2 instead of Rule 4.) And given an operation number M and a number A, an X that produces M(AX) is M52AM52. (Compare this with the former solution M32AM3.) This shows that from Rules 1 and 4 we can get both of Craig's laws. However, I have stated (Fact 2) the more general proposition that McCulloch's Law alone is enough to yield Craig's laws; this can be proved in the manner of Chapter 10—namely, given an operation number M, there is some Y that produces MY, and hence MY produces M(MY); hence X produces M(X), where X = MY. And for any A, if there is some Y that produes AMY, then MY produces M(AMY), and so X produces M(AX) for X = MY.

As for Fact 3, it can be proved just as in the last chapter. (For example, given operation numbers M and N, if Craig's Second Law holds, there is some X that produces M(N2X), and if we take N2X for Y, X produces M(Y) and Y produces N(X).)

◆ 13 ◆
The Key

Craig's affair in Norway took less time than expected, and he returned home exactly three weeks from the day he departed. When he got to his house, he found a note from McCulloch:

Dear Craig:

If, by any chance, you get back by Friday, May 12, I would very much like to have you come for dinner. I have invited Fergusson.

Best regards,

Norman McCulloch

"Excellent!" said Craig to himself. "I returned just in time!"

When Craig arrived at McCulloch's, Fergusson had already been there some fifteen minutes.

"Well, well, welcome back!" said McCulloch.

"While you were gone," said Fergusson, "McCulloch invented a new number machine!"

"Oh?" replied Craig.

"I didn't invent it all by myself," said McCulloch. "Fergusson was partly responsible. But this machine is extremely interesting; it has the following four rules:

M-I: For any number X, 2X2 produces X.

M-II: If X produces Y, then 6X produces 2Y.

M-III: If X produces Y, then 4X produces $\overleftarrow{Y}$ (as with the last machine).

M-IV: If X produces Y, then 5X produces YY (as with the last machine).

"This machine," said McCulloch, "has all the pretty properties of my last machine—it obeys your two laws and also Fergusson's double analogues."

Craig studied these rules for a while with unusual intensity.

"I can't even get off the ground," he said at last. "I can't even find a number that produces itself. Are there any?"

"Oh, yes," replied McCulloch. "Though they're much more difficult to find than they were with my last machine. In fact, I couldn't solve this problem, although Fergusson did. The shortest number we found that produces itself has ten digits."

Craig again became absorbed in thought. "Surely, the first two rules are not sufficient to yield such a number, are they?"

"Certainly not!" replied McCulloch. "One needs all four rules to get such a number."

"Remarkable!" said Craig, who then went off once more into a deep study.

"Good gracious!" he suddenly exclaimed, virtually leaping out of his chair. "Why, this solves the lock puzzle!"

"What ever are you talking about?" asked Fergusson.

"Oh, I'm sorry!" said Craig, who then told them of the entire Monte Carlo affair.

"I trust you will keep this confidential," concluded Craig, "and now, McCulloch, if you will show me a number that produces itself, I can then immediately find a combination that will open the lock."

So there are three puzzles for the reader:

(1) What number X produces itself in this latest machine?

(2) What combination will open the lock?

(3) How are the first two questions related?

EPILOGUE

Early the next morning, Craig dispatched a trusted messenger to deliver the combination to Martinez in Monte Carlo. The messenger arrived in time, and the safe was opened without incident.

True to Martinez's word, the board of directors sent Craig a handsome reward, which Craig insisted on sharing with McCulloch and Fergusson. To celebrate, the three friends spent a delightful evening at the Lion's Inn.

"Ah, yes," said Craig, after a glass of fine sherry, "this has been as interesting a case as has ever come my way! And isn't it remarkable that these number machines—invented purely out of intellectual curiosity—should have turned out to have such an unexpected practical application?"

◆ SOLUTIONS ◆

Let me first say a little more about the Monte Carlo lock puzzle.

In Farkus's last condition, nothing was said that required y to be a different combination from x. And so, taking x and y

to be the same, the condition reads: "If x is specially related to x, then if x jams the lock, x is neutral, and if x is neutral, then x jams the lock." Now, it is impossible that x can both jam the lock and be neutral; hence if x is specially related to x, then x can neither jam the lock nor be neutral; hence it must open the lock! So, if we can find a combination x that is specially related to itself, then such an x will open the lock.

Craig, of course, realized this before he came back to London. But how do you find a combination x that is specially related to itself? This is the problem Craig could not solve before he had the good fortune of seeing McCulloch's third machine.

As it turns out, the problem of finding a combination that, on the basis of Farkus's conditions, can be shown to be specially related to itself, is virtually identical with the problem of finding a number that produces itself in McCulloch's latest machine. The only essential difference is that the combinations are strings of letters, whereas the number machines work with strings of digits, but we can easily transform either problem into the other by the following simple device:

To begin with, the only combinations we need consider are those using the letters Q,L,V,R (these are obviously the only letters that play a significant role). Now suppose, instead of using these letters, we had used the respective digits 2,6,4,5 (that is, 2 for Q, 6 for L, 4 for V, 5 for R). To make this easier to remember:

Q	L	V	R
2	6	4	5

Now, let us see what Farkus's first four conditions look like when written in number notation rather than letter notation.

(1) For any number X, 2X2 is specially related to X.

(2) If X is specially related to Y, then 6X is specially related to 2Y.
(3) If X is specially related to Y, then 4X is specially related to $\overleftarrow{Y}$.
(4) If X is specially related to Y, then 5X is specially related to YY.

We at once see that these are exactly the conditions of the present number machine, except that the phrase *specially related to* is used instead of *produces.* (I could have used the term *produces* instead of *specially related to* when I presented the conditions in Chapter 8, but I did not want to give the reader too much of a hint!) And so we see how either problem can be transformed into the other.

Let me state this again, and this time more precisely: For any combination x of the letters Q,L,V,R let $\bar{x}$ be the number obtained by replacing Q by 2, L by 6, V by 4, and R by 3. For example, if x is the combination VQRLQ, then $\bar{x}$ is the number 42562. Let us call $\bar{x}$ the *code number* of x. (Incidentally, the idea of assigning numbers to expressions originated with the logician Kurt Gödel and is technically known as *Gödel numbering.* It is of great importance, as we will see in Part IV.)

Now we can precisely state the main point of the last paragraph thus: For any combination x and y of the four letters Q,L,V,R, if $\bar{x}$ can be shown to produce $\bar{y}$ on the basis of conditions M-I through M-IV of McCulloch's latest machine, then x can be shown to be specially related to y on the basis of Farkus's first four conditions—and conversely.

So, if we can find a number that must produce itself in the newest number machine, then it must be the code number of a combination that is specially related to itself, and this combination will open the lock.

Now, how do we find a number N that produces itself in

this present machine? We first look for a number H such that for any numbers X and Y, if X produces Y, then HX produces Y2Y2. If we can find such an H, then for any number Y, H2Y2 will produce Y2Y2 (because 2Y2 produces Y, by M-I), and hence H2H2 will produce H2H2, and we will have found our desired N. But how do we find such an H?

The problem boils down to this: Starting with a given number Y, how can we wind up with Y2Y2 by successively applying operations that the present machine can perform? Well, we can get Y2Y2 from Y this way: First reverse Y, getting $\overleftarrow{Y}$; then put 2 to the left of $\overleftarrow{Y}$; getting $2\overleftarrow{Y}$; then reverse $2\overleftarrow{Y}$, getting Y2; then repeat Y2, getting Y2Y2. These operations are respectively represented by the operation numbers 4, 6, 4, and 5, and so we take H to be 5464.

Let us check that this H really works: Suppose X produces Y; we are to check that 5464X produces Y2Y2. Well, since X produces Y, 4X produces $\overleftarrow{Y}$ (by M-III); hence 64X produces $2\overleftarrow{Y}$ (by M-II); hence 464X produces Y2 (by M-III); hence 5464X produces Y2Y2 (by M-IV). So if X produces Y, then HX does indeed produce Y2Y2.

Now that we have found our H, we accordingly take N to be H2H2, so the number 5464254642 produces itself (as the reader can verify directly).

Now that we know that 5464254642 produces itself, we know that it must be the code number of a combination that opens the lock. This combination is RVLVQRVLVQ.

Of course, the Monte Carlo lock problem can be solved directly, rather than by translating it into a number-machine problem, but I did the latter because, for one thing, it happened historically that this is the way Craig found the solution. For another thing, I felt it would be of interest for the reader to see an example of how two mathematical problems can have different contents but the same abstract form.

To verify directly that RVLVQRVLVQ is specially related to itself (and hence opens the lock), we reason as follows: QRVLVQ is specially related to RVLV (by Property Q); hence VQRVLVQ is specially related to the reverse of RVLV (by Property V), which is VLVR. Therefore, LVQRVLVQ is specially related to QVLVR (by Property L), and hence VLVQRVLVQ is specially related to the reverse of QVLVR, which is RVLVQ. Hence (by Property R), RVLVQRVLVQ is specially related to the repeat of RVLVQ, which is RVLVQRVLVQ. And so RVLVQRVLVQ is specially related to itself.

PART FOUR

SOLVABLE OR UNSOLVABLE?

◆ 14 ◆
Fergusson's Logic Machine

Some months after the celebrated solution of the Monte Carlo lock mystery, Craig and McCulloch paid a visit to Fergusson to learn about his logic machine. It did not take long for the conversation to turn to the nature of provability.

"I must tell you an interesting and revealing incident," said Fergusson. "A student was asked on a geometry examination to prove the Pythagorean theorem. He handed in his paper, and the Mathematics-Master returned it with a grade of zero and the comment, 'This is no proof!' Later, the lad went to the Mathematics-Master and said, 'Sir, how can you say that what I handed you is not a proof? You have never once in this course defined what a proof is! You have been admirably precise in your definitions of such things as triangles, squares, circles, parallelity, perpendicularity, and other geometric notions, but never once have you defined exactly what you mean by the word 'proof.' How, then, can you so assuredly assert that what I have handed you is not a proof? How would you *prove* that it is not a proof?' "

"Brilliant!" exclaimed Craig, clapping his hands. "That boy will go far. How did the Master respond?"

"Oh," replied Fergusson, "unfortunately the Master was a dry pedantic sort with no sense of humor and no imagination.

He took off additional marks on the grounds that the boy was being impertinent."

"How unfortunate!" exclaimed Craig in indignation. "Had I been the Master, I would have given the boy highest honors for such a keen observation!"

"Of course," replied Fergusson. "So would I. But you know how it is with unfortunately too many teachers; they have no creative ability of their own and feel threatened by students who can think for themselves."

"I must admit," said McCulloch, "that if I had been in the Master's place, I also could not have answered the boy's question. Of course, I would have complimented him for raising the question, but I don't see how I could have answered it. Just what is a proof, anyhow? I somehow seem to recognize a correct proof when I see one, and I can usually spot an invalid argument when I come across one; still, if asked for a *definition* of a proof, I would be sorely pressed!"

"That's the situation with almost all working mathematicians," replied Fergusson. "More than ninety-nine percent of them can recognize a correct proof or spot an invalidity in an incorrect proof, even though they cannot define what they mean by a proof. One task we logicians are interested in is that of analyzing the notion of 'proof'—to make it as rigorous as any other notion in mathematics."

"If most mathematicians already know what a proof is, even though they can't define one," said Craig, "why is it so important that the notion be defined?"

"There are several reasons," replied Fergusson. "Although even if there were none, I would like to know the definition for its own sake. It has frequently happened in the history of mathematics that certain basic notions—for example, continuity—were intuitively apprehended long before they were rigorously defined. Once defined, however, the notion acquires a new dimension; facts about it can be established that would be exceedingly difficult, if not impossible, to discover without a firm criterion of when the notion does or does not apply. The notion of 'proof' is no exception; it has sometimes happened that a proof utilizes a new principle—such as the Axiom of Choice—and that controversies arise as to whether the principle is legitimate. A precise definition of 'proof' pinpoints just what mathematical principles are or are not being used.

"For another thing, it becomes particularly critical to have a precise notion of 'proof' when one wishes to establish that a given mathematical statement is *not* provable from a given set of axioms. The situation is analogous to ruler-and-compass constructions in Euclidean geometry: to show that a certain construction, such as trisecting an angle, squaring a circle, or constructing a cube with twice the volume of a given cube, is *not* possible involves a more critical characterization of the notion of 'construction' than does a positive result in the form that such-and-such a construction *is* possible with ruler and compass. And so it is with provability: to show that a given statement is *not* provable from a given set of axioms requires a more critical characterization of the notion of *proof* than a positive result in the form that a given statement *is* provable from the axiom."

A GÖDELIAN PUZZLE

"Now," continued Fergusson, "given an axiom system, a proof in the system consists of a finite sequence of sentences constructed according to very precise rules. It is a simple matter to decide purely mechanically whether a given sequence of sentences is or is not a proof in the system; indeed, it is a simple matter to construct a machine that does this. It is an altogether different matter to construct a machine that will decide which sentences of an axiom system are provable and which ones are not. Whether or not this can be done may, I suspect, depend on the axiom system. . . .

"My current interest is in mechanical theorem-proving— that is, in machines that prove various mathematical truths. Here is my latest one," Fergusson said, pointing proudly to an extremely odd-looking contraption.

Craig and McCulloch stood several minutes before the machine trying to figure out its functions.

"Just what does it do?" Craig finally asked.

"It proves various facts about the positive whole numbers," replied Fergusson. "I am working in a language that contains names of various sets of numbers—specifically, positive integers. There are infinitely many sets of numbers nameable in this language. For example, we have a name for the set of even numbers, one for the set of odd numbers, one for the set of prime numbers, one for the set of all numbers divisible by 3—just about every set that number-theorists are interested in has a name in the language. Now, although there are infinitely many nameable sets, there are no more nameable sets than there are positive integers. And to each positive integer n is associated a certain nameable set A_n. We can thus think of all the nameable sets arranged in an infinite sequence $A_1, A_2, \ldots, A_n. \ldots$ (If you like, you can think of a book with infinitely many pages, and for each positive integer n, the nth page contains a description of a set of positive integers. Then think of the set A_n as the set described on page n of the book.)

"I employ the mathematical symbol 'ϵ,' which represents the English phrase 'belongs to' or 'is a member of,' and for every number x and every number y, we have the sentence $x \in A_y$, which is read 'x belongs to the set A_y.' This is the only type of sentence my machine investigates; the function of the machine is to try and discover what numbers belong to what nameable sets.

"Now, each sentence $x \in A_y$ has a *code* number—namely, a number which, when written in the usual base 10 notation, consists of a string of 1's of length x followed by a string of 0's of length y. For example, the code number of the sentence $3 \in A_2$ is 11100; the code number of $1 \in A_5$ is 100000. For any x and y, by $x{*}y$ I mean the code number of the sentence $x \in A_y$; thus, $x{*}y$ consists of a string of 1's of length x followed by a string of 0's of length y.

"The machine operates in the following manner," continued Fergusson. "Whenever it discovers that a number x belongs to a set A_y, it then prints out the number $x{*}y$—the code number of the sentence $x \in A_y$. If the machine prints $x{*}y$, then I say that the machine has *proved* the sentence $x \in A_y$. And I say that the sentence $x \in A_y$ is *provable* (by the machine) if the machine is capable of printing out the number $x{*}y$.

"Now, I know that my machine is always accurate in the sense that every sentence provable by the machine is true."

"Just a moment," interrupted Craig, "what do you mean by *true*? How does *true* differ from *provable*?"

"Oh," replied Fergusson, "the two concepts are entirely different: I call a sentence $x \in A_y$ *true* if x is really a member of the set A_y. That is entirely different from saying that the machine is capable of printing out the number $x{*}y$. If the latter holds, then I say that the sentence $x \in A_y$ is *provable*—that is, by the machine."

"Oh, now I understand," said Craig. "In other words, when

you say that your machine is accurate—that every sentence provable by the machine is a true sentence—what you mean is that the machine never prints out a number $x*y$ unless x is really a member of the set A_y. Is that correct?"

"Exactly!" replied Fergusson.

"Tell me," said Craig, "how do you know that your machine is always accurate?"

"To answer that," replied Fergusson, "I must tell you all the details of the machine. The machine operates on the basis of certain axioms about the positive integer; these axioms have been programmed into the machine in the form of certain instructions. The axioms are all well-known mathematical truths. The machine cannot prove any statement that is not a logical consequence of the axioms. Since the axioms are all true, and any logical consequence of true statements must be true, then the machine is incapable of proving a false sentence. I can tell you the axioms if you like, and then you can see for yourselves that the machine can prove only true sentences."

"Before you do that," said McCulloch, "I would like to ask another question. Suppose I am willing temporarily to take your word that every sentence provable by the machine is true. What about the converse? Is every true sentence of the form $x \in A_y$ provable by the machine? In other words, is the machine capable of proving *all* true sentences of the form $x \in A_y$, or only some?"

"A most important question," replied Fergusson, "but, alas, I don't know the answer! That is precisely the basic problem I have been unable to solve! I have been working on it on and off for months but have gotten nowhere. I know for sure that the machine can prove every statement $x \in A_y$ that is a logical consequence of the axioms, but I don't know whether I have programmed in enough axioms. The axioms in question represent just about the sum total of what mathematicians know about the system of positive integers; still, there may not be enough to settle completely which numbers x belong to which nameable sets A_y. So far, every sentence $x \in A_y$ that I have examined and found to be true on purely mathematical grounds I have found to be a logical consequence of the axioms, and so the machine is capable of proving it. But just because I have not yet been able to find a true sentence that the machine cannot prove doesn't mean that there isn't one; it might be that I just haven't found it. Or, then again, it may be that the machine *can* prove all true sentences; but I have not yet been able to prove this fact. I just don't know!"

To make a long story short, at this point Fergusson told Craig and McCulloch all the axioms used by the machine, as well as the purely logical rules that enabled it to prove new sentences from old ones. Once Craig and McCulloch knew these details of the machine's operation, they could see immediately that it was indeed accurate—that it did prove only true sentences. But this still left unsolved the problem of whether the machine could prove all true sentences or only some. The three met together several times during the next few months and slowly but surely closed in on the problem, until they finally solved it.

I will not burden the reader with all the details, but will mention only those that are relevant to the solution of the problem. The turning point in the investigation came when the three men worked out three key properties of the machine; these properties sufficed to settle the question. It was, I believe, Craig and McCulloch who first brought the three properties to light, but it was Fergusson who applied the finishing touches. I will tell you what these properties are in a moment; but first, a little preliminary notation.

For any set A of positive integers, by its *complement* $\bar{A}$ is meant the set of all positive integers that are not in A. (For example, if A is the set of even numbers, then its complement $\bar{A}$ is the set of odd numbers; if A is the set of numbers divisible by 5, then its complement $\bar{A}$ is the set of numbers that are not divisible by 5.)

For any set A of positive integers, by A^* we shall mean the set of all positive integers x such that $x{*}x$ is a member of A. Thus, for any number x, to say that x lies in A^* is equivalent to saying that $x{*}x$ lies in A.

Now, here are the three key properties that Craig and McCulloch discovered about the machine:

Property 1: The set A_8 is the set of all numbers that the machine is capable of printing.

Property 2: For each positive integer n, $A_{3 \cdot n}$ is the complement of A_n. (By $3 \cdot n$ we mean 3 times n.)

Property 3: For every positive integer n, the set $A_{3 \cdot n+1}$ is the set $A_n{}^*$ (the set of all numbers x such that $x{*}x$ belong to A_n).

1

From Properties 1, 2, and 3, it can be rigorously deduced that Fergusson's machine is *not* able to prove all true sentences! The problem for the reader is to find a sentence that is true but not provable by the machine. That is, we are to find numbers n and m (either the same or different) such that n is in fact a member of the set A_m, yet the code number $n{*}m$ of the sentence $n \in A_m$ cannot possibly be printed by the machine.

2

In the solution given for Problem 1, the numbers *n* and *m* were both less than 100. There is another such solution in which *n* and *m* are both less than 100 (and again, *m* might be the same as *n* or different; I'm not telling which). Can the reader find it?

3

Without any restriction on the sizes of *n* and *m*, how many solutions are there? That is, how many sentences are there which are true but not provable by Fergusson's machine?

EPILOGUE

Fergusson did not easily give up his aspiration of constructing a machine that could prove all arithmetic truths without proving any falsehoods, and he constructed many, many more logic machines.* But for each machine he constructed, either he or Craig or McCulloch discovered a true sentence that the machine could not prove. And so he finally gave up the attempt to construct a purely mechanical device that was both accurate and could prove all true arithmetic sentences.

That Fergusson's heroic attempt failed was not due to any lack of ingenuity on his part. We must remember that he lived several decades before the discoveries of such logicians as Gödel, Tarski, Kleene, Turing, Post, Church, and others, to

* Some of them were quite interesting, and I hope to tell you about them in another book.

whose work we will soon turn. Had he lived to see what these men produced, he would have realized that his failure stemmed exclusively from the fact that what he was attempting was inherently impossible! And so, with a salute to Fergusson, and his colleagues Craig and McCulloch, we shall jump ahead three or four decades and take a look at the critical year 1931.

• SOLUTIONS •

1 • One solution is that the sentence $75 \epsilon A_{75}$ is true but not provable by the machine. Here is the reason why.

Suppose the sentence $75 \epsilon A_{75}$ were false. Then 75 doesn't belong to the set A_{75}. Hence 75 must belong to A_{25} (by Property 2, which makes A_{75} the complement of A_{25}). This implies (by Property 3) that $75*75$ belongs to A_8, since $25 = 3\cdot 8 + 1$; and hence that $75*75$ can be printed by the machine; in other words, that $75 \epsilon A_{75}$ is provable by the machine. Thus, if the sentence $75 \epsilon A_{75}$ were false, it would be provable by the machine. But we are given that the machine is accurate and never proves false sentences. Therefore, the sentence $75 \epsilon A_{75}$ cannot be false; it must be true.

Since the sentence $75 \epsilon A_{75}$ is true, then 75 does belong to the set A_{75}. Hence 75 cannot belong to the set A_{25} (by Property 2), and hence the number $75*75$ cannot belong to A_8, because if $75*75$ did belong to A_8, then, by Property 3, 75 would belong to A_{25}. Since $75*75$ doesn't belong to A_8, then the sentences $75 \epsilon A_{75}$ is not provable by the machine. And so the sentence $75 \epsilon A_{75}$ is true but not provable by the machine.

2 • Before giving other solutions, let us observe the following general fact: The key set K is the set of all numbers x such

that the sentence $x \epsilon A_x$ is not provable by the machine—or what is the same thing, the set of all numbers x such that $x*x$ cannot be printed by the machine. Now, A_{75} is this set K, because to say that x belongs to A_{75} is equivalent to saying that x does not belong to A_{25}, which in turn is equivalent to saying that $x*x$ does not belong to A_8, which is the set of all numbers that the machine *can* print. So $A_{75} = K$. But also $A_{73} = K$, because to say that a number x belongs to A_{73} is equivalent to saying that $x*x$ belongs to A_{24} (by Property 3, since $73 = 3 \cdot 24 + 1$), which in turn is equivalent to saying that $x*x$ does not belong to A_8 (by Property 2). Thus, A_{73} is the set of all numbers x such that $x*x$ does not belong to A_8—or, what is the same thing, such that $x \epsilon A_x$ is not provable by the machine. Thus, A_{73} is the same set as A_{75}, since both are the same as the set K. Moreover, given *any* number n such that $A_n = K$, the sentence $n \epsilon A_n$ must be true but not provable by the machine—this by essentially the same argument as for the special case $n = 75$ (an argument we give in a still more general form in the next chapter). And so $73 \epsilon A_{73}$ is another example of a true sentence whose code number the machine cannot print.

3 • For any n, the set $A_{9 \cdot n}$ must be the same as the set A_n, because $A_{9 \cdot n}$ is the complement of $A_{3 \cdot n}$, and $A_{3 \cdot n}$ is the complement of A_n; hence $A_{9 \cdot n}$ is the same set as A_n. And so A_{675} is the same as the set A_{75}, and so $675 \epsilon A_{675}$ is another solution. Also $2175 \epsilon A_{2175}$ is a solution. Indeed, there are infinitely many true sentences that Fergusson's machine cannot prove: for any n that is 75 times some multiple of 9, or 73 times some multiple of 9, the sentence $n \epsilon A_n$ is true but not provable by the machine.

◆ 15 ◆
Provability and Truth

The year 1931 was indeed a great landmark in the history of mathematical logic; this was the year in which Kurt Gödel published his famous Incompleteness Theorem. Gödel begins his epoch-making paper* as follows:

> The development of mathematics in the direction of greater precision has led to large areas of it being formalized, so that proofs can be carried out according to a few mechanical rules. The most comprehensive formal systems to date are, on the one hand, the *Principia Mathematica* of Whitehead and Russell and, on the other, the Zermelo-Fraenkel system of axiomatic set theory. Both systems are so extensive that all methods of proof used in mathematics today can be formalized in them; i.e., can be reduced to a few axioms and rules of inference. It would seem reasonable, therefore, to surmise that these axioms and rules of inference are sufficient to decide *all* mathematical questions which can be

* "Über formal unentscheidbare Sätze der *Principia Mathematica* und verwandter Systeme I" ("On Formally Undecidable Propositions of *Principia Mathematica* and Related Systems"), *Monatshefte für Mathematik und Physik* 38, 173–198.

> formulated in the system concerned. In what follows it will be shown that this is not the case, but rather that, in both of the cited systems, there exist relatively simple problems of the theory of ordinary whole numbers which cannot be decided on the basis of the axioms.†

Gödel then goes on to explain that the situation does not depend on the special nature of the two systems under consideration, but holds for an extensive variety of mathematical systems.

Just what is this "extensive variety" of mathematical systems? Various interpretations have been given, and Gödel's theorem has accordingly been generalized in several ways. Curiously enough, one of the ways that is most direct and most easily accessible to the general reader is also the way that appears to be the least well known. What makes this even more curious is that the way in question is the very one indicated by Gödel himself in the introductory section of his original paper! To which we shall now turn.

Let us consider an axiom system with the following properties: First of all, we have names for various sets of (positive whole) numbers, and we have all these nameable sets arranged in an infinite sequence $A_1, A_2, \ldots, A_n, \ldots$ (just as in Fergusson's system of the last chapter). We shall call a number n an *index* of a nameable set A if $A = A_n$. (Thus, for example, if the sets A_2, A_7, and A_{13} all happen to be the same, then 2, 7, and 13 are all indices of this set.) As with Fergusson's system, we have associated with every number x and every number y a certain sentence—written $x \in A_y$—which is called *true* if x belongs to A_y and *false* if x doesn't belong to A_y. We no longer assume, however, that the sentences $x \in A_y$ are the only sentences of the system; there may be others. But

† Composite translation.

every other such sentence is classified as either a true sentence or a false one.

Every sentence of the system is assigned a code number, which we will now call its *Gödel number,* and we will let $x*y$ be the Gödel number of the sentence $x \epsilon A_y$. (We no longer need assume that $x*y$ consists of a string of 1's of length x followed by a string of 0's of length y; this is nothing at all like the code numbering that Gödel actually used. There are many different code numberings that work, and which one works most smoothly depends on the particular system under consideration. Anyway, for the general theorem we are about to prove, nothing about the particular Gödel numbering need be assumed.)

Certain sentences are taken as *axioms* of the system, and certain rules are provided that enable one to prove various sentences from the axioms. There is, thus, a well-defined property of a sentence being *provable* in the system. It is assumed that the system is *correct,* in the sense that every sentence provable in the system is true; hence, in particular, that whenever a sentence $x \epsilon A_y$ is provable in the system, then x really is a member of the set A_y.

We let P be the set of Gödel numbers of all the sentences provable in the system. For any set A of numbers, we again let $\bar{A}$ be the complement of A (the set of all numbers not in A) and we let A^* be the set of all numbers x such that $x*x$ belongs to A. We are now interested in systems in which the following conditions, G1, G2, and G3, hold:

G1: The set P is nameable in the system. Stated otherwise, there is at least one number p such that A_p is the set of Gödel numbers of the provable sentences. (For Fergusson's system, 8 was such a number p.)

G2: The complement of any set nameable in the system is also nameable in the system. Stated otherwise, for any number x there is some number x' such that $A_{x'}$ is the complement of A_x. (For Fergusson's system, $3{\cdot}x$ was such a number x'.)

G3: For any nameable set A, the set A^* is also nameable in the system. Stated otherwise, for any number x there is some number x^* such that A_{x^*} is the set of all numbers n such that $n{*}n$ lies in A_x. (For Fergusson's system, $3{\cdot}x + 1$ was such a number x^*.)

The conditions F1, F2, and F3 characterizing Fergusson's machine are obviously nothing more than special cases of conditions G1, G2, and G3. These latter general conditions are of considerable importance, because they do hold for a wide variety of mathematical systems, including the two systems treated in Gödel's paper. That is, it is possible to arrange all the nameable sets in an infinite sequence $A_1, A_2, \ldots, A_n, \ldots$ and to exhibit a particular Gödel numbering of the sentences such that the conditions G1, G2, and G3 do hold. Therefore, anything provable about systems satisfying the conditions G1, G2, and G3 will apply to many important systems.

We can now state and prove the following abstract form of Gödel's theorem.

Theorem G: Given any correct system satisfying conditions G1, G2, and G3, there must be a sentence that is true but not provable in the system.

The proof of Theorem G is a straightforward generalization of the proof that the reader already knows for Fergusson's system: We let K be the set of all numbers x such that $x{*}x$ is not in the set P. Since P is nameable (by G1), so is its complement $\bar{P}$ (by G_2), and hence so is the set $\bar{P}^*$ (by G_3); but $\bar{P}^*$ is the set K (because $\bar{P}^*$ is the set of all numbers x such that $x{*}x$ lies in $\bar{P}$, or, what is the same thing, the set of all x such that $x{*}x$ doesn't lie in P). And so the set K is nameable in the system, which means that $K = A_k$ for at least one number

k. (For Fergusson's system, 73 was one such number k, and so was 75.) Thus, for any number x, the truth of the sentence $x \in A_k$ is tantamount to the assertion that $x*x$ is not in P, which is tantamount to the fact that the sentence $x \in A_x$ is not provable (in the system). In particular, if we take k for x, the truth of the sentence $k \in A_k$ is tantamount to its nonprovability in the system, which means that it is either true and not provable in the system or false but provable in the system. The latter possibility is out, since we are given that the system is correct; hence it must be that the form holds—the sentence is true but not provable in the system.

Discussion: In *What Is the Name of This Book?* I considered the analogous situation of an island on which every inhabitant is either a *knight* who always tells the truth or a *knave* who always lies. Certain knights were called *established* knights and certain knaves were called *established* knaves. (The knights correspond to *true* sentences, and the established knights correspond to sentences that are not only true but *provable.*) Now, it is impossible for any inhabitant of the island to say, "I am not a knight," because a knight would never lie and claim he wasn't a knight, and a knave would never truthfully admit to not being a knight. Therefore, no inhabitant of the island can claim that he is not a knight. However, it *is* possible for an inhabitant to say, "I am not an established knight." If he says that, no contradiction arises, but something interesting follows—namely, that the speaker must *in fact* be a knight but not an established knight. For a knave would never make the truthful claim that he is not an established knight (for in truth he isn't), and so the speaker must be a knight. Since he is a knight, his statement must be true; so he is a knight but, as he says, not an established knight—just as the sentence $k \in A_k$, which asserts its own nonprovability in the system, must be true but not provable in the system.

GÖDEL SENTENCES AND TARSKI'S THEOREM

Let us now consider a system satisfying at least conditions G2, G3 (for the time being, condition G1 is not relevant). We have defined P to be the set of Gödel numbers of the provable sentences of the system; let us now define T to be the set of Gödel numbers of all the *true* sentences of the system. In the year 1933, the logician Alfred Tarski raised and answered the following question: Is the set T nameable in the system, or not? This question can be answered purely on the basis of conditions G2 and G3. I will give the answer shortly, but first let's turn to a still more basic question of systems that satisfy at least the condition G3.

Given any sentence X and any set A of positive whole numbers, we shall call X a *Gödel sentence* for A if either X is true and its Gödel number lies in A, or X is false and its Gödel number lies outside A. (Such a sentence can be thought of as asserting that its own Gödel number lies in A; if the sentence is true, then its Gödel number really is in A; if the sentence is false, then its Gödel number is not in A.) Now, we shall call a system *Gödelian* if for every set A nameable in the system there is at least one Gödel sentence for A.

The following fact is basic:

Theorem C: If a system satisfies condition G3, then it is Gödelian.

1

Prove Theorem C.

2

To take a special case, consider Fergusson's system. Find a Gödel sentence for the set A_{100}.

3

Suppose a system is Gödelian (without necessarily satisfying condition G3). If the system is correct and satisfies conditions G1 and G2, does it necessarily contain a sentence that is true but not provable in the system?

4

Let T be the set of Gödel numbers of the true sentences. Is there a Gödel sentence for T? Is there a Gödel sentence for $\bar{T}$, the complement of T?

Now we are in a good position to give the answer to Tarski's question. The following is an abstract version of Tarski's theorem.

Theorem T: Given any system satisfying conditions G2 and G3, the set T of Gödel numbers of the true sentence is *not* nameable in the system.

Note: The word *definable* is sometimes used in place of *nameable*, and Theorem T is sometimes paraphrased as follows: For sufficiently rich systems, truth within the system is not definable within the system.

5

Prove Theorem T.

6

It is instructive to note that once Theorem T has been proved, one can immediately obtain Theorem G as a corollary. Can the reader see how?

A DUAL FORM OF GÖDEL'S ARGUMENT

The various systems that have been proved incomplete by Gödel's argument also have the property that associated with each sentence X is a sentence X′ called the ***negation*** of X, which is true if and only if X is false. A sentence X is called ***disprovable*** or ***refutable*** in the system if its negation X′ is provable in the system. Assuming the system to be correct, no false sentence is provable in the system and no true sentence is refutable in the system.

We have seen that conditions G1, G2, G3 imply the existence of a Gödel sentence G for the set $\bar{P}$, and that such a sentence G is true but not provable in the system (assuming the system is correct). Since G is true, it can't be refutable in the system either (again by the assumption of correctness). So the sentence G is neither provable nor refutable in the system. (Such a sentence is called ***undecidable*** in the system.)

In a 1960 monograph, "Theory of Formal Systems," I considered a "dual" form of Gödel's argument: Instead of a sentence that asserts its own nonprovability, what about constructing a sentence that asserts its own refutability? More precisely, let R be the set of Gödel numbers of the refutable sentences. Suppose X is a Gödel sentence for R; what is the status of X? This idea is carried out in the next problem.

7

Let us now consider a correct system that satisfies condition G3, but instead of assuming conditions G1, G2, we assume the following single condition:

G1′: The set R is nameable in the system.

(Thus we assume that the system is correct and satisfies conditions G1′ and G3.)

(a) Prove that there is a sentence which is neither provable nor refutable in the system.

(b) To take a special case, suppose we are given that A_{10} is the set R and that for any number n, A_{5*n} is the set of all x such that $x*x$ is in A_n (this is a special case of G3). The problem now is actually to find a sentence that is neither provable nor refutable in the system, and to determine whether the sentence is true or false.

Remarks: (1) Gödel's method of obtaining an undecidable sentence boils down to constructing a Gödel sentence for $\bar{P}$, the complement of P; such a sentence (which can be thought of as asserting its own nonprovability) must be true but not provable in the system. The "dual" method boils down to constructing a Gödel sentence for the set R rather than for the set $\bar{P}$; such a sentence (which can be thought of as asserting its own refutability) must be false but not refutable. (Since it is false, it is not provable either, hence is undecidable in the system.) I should remark that the systems treated in Gödel's original paper satisfy all four conditions, G1, G2, G3, and G1′, so either method can be used for constructing undecidable sentences.

(2) Just as a sentence that asserts its nonprovability is like a native of a knight-knave island who claims that he is not an established knight, so a sentence that asserts its own refutability is like a native of the island who claims that he is an established *knave;* such a native is indeed a knave, but not an established one. (I leave the proof of this to the reader.)

◆ SOLUTIONS ◆

1 ◆ Suppose the system does satisfy condition G3. Let S be any set nameable in the system. Then, by G3, the set S^* is nameable in the system. So there is some number b such that $A_b = S^*$. Now, a number x belongs to S^* just in case $x*x$ belongs to S. So a number x belongs to A_b just in case $x*x$ belongs to S. In particular, taking b for x, the number b belongs to A_b just in case $b*b$ belongs to S. Also, b belongs to A_b if and only if the sentence $b \in A_b$ is true. So $b \in A_b$ is true if and only if $b*b$ belongs to S. Also, $b*b$ is the Gödel number of the sentence $b \in A_b$. And so we see that $b \in A_b$ is true if and only if its Gödel number belongs to S. So if $b \in A_b$ is true, its Gödel number belongs to S; if $b \in A_b$ is false, its Gödel number does not belong to S. Thus, the sentence $b \in A_b$ is a Gödel sentence for S.

2 ◆ In Fergusson's system, given any number n, $A_{3 \cdot n+1}$ is the set A_n^*. And so A_{301} is the set A_{100}^*. And so we use the result of the last problem and take 301 for b. Thus, $301 \in A_{301}$ is a Gödel sentence for the set A_{100}. More generally, for *any* num-

ber n, if we let $b = 3 \cdot n+1$, the sentence $b \in A_b$ is a Gödel sentence for A_n in Fergusson's system.

3 ◆ Yes, it does: Suppose the system is Gödelian and conditions G1 and G2 both hold, and suppose also that the system is correct. By G1 the set P is nameable; hence, by G2, $\bar{P}$, the complement of P, is nameable. Then, since the system is Gödelian, there is a Gödel sentence X for $\bar{P}$. This means that X is true if and only if X's Gödel number is in $\bar{P}$. But to say that X's Gödel number is in $\bar{P}$ is to say that it is not in P, which is the same thing as saying that X is not provable. Thus, a Gödel sentence for $\bar{P}$ is nothing more nor less than a sentence that is true if and only if it is not provable in the system, and (as we have seen) such a sentence must be true but not provable in the system (assuming the system is correct).

Indeed, the essence of Gödel's argument is the construction of a Gödel sentence for the set $\bar{P}$.

4 ◆ Obviously, every sentence X is a Gödel sentence for T, because if X is true its Gödel number is in T, and if X is false its Gödel number is not in T. Therefore, *no* sentence can be a Gödel sentence for $\bar{T}$, because it cannot be that either X is true and its Gödel number is in $\bar{T}$, or that X is false and its Gödel number is not in $\bar{T}$.

It might be instructive for the reader to observe that for *any* number set A and for any sentence X, X is either a Gödel sentence for A or X is a Gödel sentence for $\bar{A}$, but never both.

5 ◆ Let us first consider any system satisfying condition G3. By Problem 1, given any set nameable in the system, there is a Gödel sentence for it. Also, by the last problem, there is no Gödel sentence for the set $\bar{T}$. Therefore, if the system satisfies

G3, the set $\bar{T}$ is not nameable in the system. If the system also satisfies condition G2, then T is not nameable in the system either—because if it were, then, by G2, so would its complement $\bar{T}$ be nameable, which it isn't. This proves that in a system satisfying G2 and G3, the set T is not nameable in the system.

In summary: (a) If G3 holds, then $\bar{T}$ is not nameable; (b) if G2 and G3 both hold, then neither T nor $\bar{T}$ is nameable in the system.

6 • If we have first proved Theorem T, we can obtain Theorem G as follows:

Suppose we have a correct system satisfying conditions G1, G2, G3. From G2 and G3, using Theorem T, we see that T is not nameable in the system. But, by G1, P *is* nameable in the system. Since P is nameable and T is not, then P and T must be different sets. However, every number in P is also in T, since we are given that the system is correct in that every provable sentence is true. Therefore, since T is not the same as the set P, there must be at least one number n in T that is not in P. Since n is in T, it must be the Gödel number of a sentence X which is true. But since n is not in P, then X is not provable in the system. Thus, X is true but not provable in the system. Thus Theorem G holds.

7 • We are given conditions G1′ and G3.

(a) By G1′, the set R is nameable in the system. Then, by condition G3, the set R^* is nameable in the system. Hence, there is some number h such that $A_h = R^*$. Now, by the definition of R^*, a number x is in R^* if and only if $x*x$ is in R. Therefore, for any x, x belongs to A_h if and only if $x*x$ belongs to R. In particular, if we take h for x, h belongs to A_h if and only if $h*h$ belongs to R. Now, h belongs to A_h if and only if the sentence $h \in A_h$ is true. Also, since $h{*}h$ is the Gödel number of the sentence $h \in A_h$, then $h{*}h$ belongs to R if and only if the sentence $h \in A_h$ is refutable. Therefore, the sentence $h \in A_h$ is true if and only if it is refutable. This means that the sentence is either true and refutable or false and not refutable. It cannot be true and refutable, since we are given that the machine is correct; hence it must be false but not refutable. Since the sentence is false, it cannot be provable either (again, because the system is correct). Therefore, the sentence $h \in A_h$ is neither provable nor refutable (and, also, it is false).

(b) We are now given that A_{10} is R, and also that for any n, $A_{5{\cdot}n}$ is the set A_n^*. Therefore, A_{50} is the set R^*. And so, by Solution (a), taking 50 for h, the sentence $50 \in A_{50}$ is neither provable nor refutable. Also, the sentence is false.

◆ 16 ◆
Machines That Talk About Themselves

We shall now consider Gödel's argument from a slightly different perspective, which puts the central idea in a remarkably clear light.

We shall take the four symbols P,N,A, – and consider all possible combinations of these symbols. By an *expression* we mean any combination of the symbols. For example, P – – NA – P is an expression; so is – PN – – A – P – . Certain expressions will be assigned a meaning, and these expressions will be called *sentences.*

Suppose we have a machine that can print out some expressions but not others. We call an expression *printable* if the machine can print it. We assume that any expression that the machine can print will be printed sooner or later. Given any expression X, if we wish to express the proposition that X is printable, we write P – X. So, for example, P – ANN says that ANN is printable (this may be true or false, but that's what it says!). If we want to say that X is *not* printable, we write NP – X. (The symbol N is the abbreviation of the word *not,* just as the symbol P represents the word *printable.* And so NP – X is to be read, crudely, as "Not printable X," or, in better English, "X is not printable.")

By the *associate* of an expression X we mean the expression X – X. We use the symbol A to stand for "the associate of," and so, for any given X, if we wish to state that the associate of X is printable we write PA – X (read "printable the associate of X," or in better English, "the associate of X is printable"). If we wish to say that the associate of X is *not* printable, we write NPA – X (read "not printable the associate of X," or, in better English, "the associate of X is not printable").

Now, the reader may well wonder why we use the dash as a symbol: why don't we simply use PX rather than P – X to express the proposition that X is printable? The reason is that omission of the dash would create a contextual ambiguity. What, for example, would PAN mean? Would it mean that the associate of N is printable or that the expression AN is printable? With the use of the dash, no such ambiguity arises. If we want to say that the associate of N is printable, we write down PA – N; whereas, if we want to say that AN is printable, we write down P – AN. Again, suppose we want to say that – X is printable; do we write P – X? No, that would state that X is printable. To say that – X is printable, we must write P – – X.

Perhaps some more examples might help: P – – says that – is printable; PA – – says that – – – (the associate of –) is printable; P – – – – also says that – – – is printable; NPA – – P – A says that the associate of – P – A is not printable; in other words, that – P – A – – P – A is not printable. NP – – P – A – – P – A says the same thing.

We now define a *sentence* as any expression of one of the four forms P – X, NP – X, PA – X, and NPA – X, where X is any expression whatever. We call P – X *true* if X is printable, and *false* if X is not printable. We call NP – X true if X is *not* printable and false if X *is* printable. We call PA – X true if the associate of X is printable, and false if the associate of X is not printable. Finally, we call NA – X true if the associate of X is not printable, and false if the associate of X is printable.

We have now given a precise definition of truth and falsity for sentences of all four types, and from this it follows that, for any expression X:

Law 1: P – X is true if and only if X is printable (by the machine).

Law 2: PA – X is true if and only if X – X is printable.

Law 3: NP – X is true if and only if X is not printable.

Law 4: NPA – X is true if and only if X – X is not printable.

We have here a curious loop! The machine is printing out sentences that make assertions about what the machine can and cannot print! In this sense, the machine is talking about itself (or, more accurately, printing out sentences about itself).

We are now given that the machine is a hundred percent accurate—that is, it never prints out any false sentence; it prints out only true sentences. This fact has several ramifications: As an example, if it ever prints out P – X, then it must also print out X, because, since it prints out P – X, then P – X must be true, which means that X is printable, and hence the machine must sooner or later print X.

It follows as well that if the machine should print out PA – X, then (since PA – X must be true), the machine must also print out X – X. In addition, if the machine prints out NP – X, then it *cannot* also print P – X, since these two sentences can't both be true—the first says that the machine doesn't print X, and the second says that the machine does print X.

The following problem puts Gödel's idea into as clear a light as any problem I can imagine.

1 • A Singularly Gödelian Challenge

Find a true sentence that the machine cannot print!

2 • A Doubly Gödelian Puzzle

We continue to assume the same conditions—and, in particular, that the machine is accurate.

There is a sentence X and a sentence Y such that one of the sentences X or Y must be true but not printable, but it is impossible to tell, from the given conditions embodied in Laws 1 through 4, which one it is. Can you find such an X and Y? (Hint: Find sentences X and Y such that X says that Y is printable and Y says that X is not printable. There are two different ways of doing this, and both relate to Fergusson's laws!)

3 • A Triply Gödelian Problem

Construct sentences X, Y, and Z such that X says that Y is printable, Y says that Z is not printable, and Z says that X is printable, and show that at least one of these three sentences (though it can't be determined which) must be true but not printable by the machine.

TWO MACHINES THAT TALK ABOUT THEMSELVES AND ABOUT EACH OTHER

Let us now add a fifth symbol, R. We thus have the symbols P,R,N,A, –, and are now given two machines M1 and M2, each of which prints out various expressions composed of these five symbols. We now interpret "P" to mean "printable by the first machine," and we interpret "R" to mean "printable by the second machine." Thus, P – X now means that X is printable by the first machine, and R – X means that X is printable by the second machine. Also, PA – X means that the associate of X is printable by the first machine, RA – X that the associate of X is printable by the second machine. Also, NP – X, NR – X, NPA – X, NRA – X respectively mean: X is not printable by the first machine; X is not printable by the second machine; X – X is not printable by the first machine; X – X is not printable by the second machine. By a *sentence* is now meant any expression of one of the eight types P – X, R – X, NP – X, NR – X, PA – X, RA – X, NPA – X, or NRA – X, and we are given that the first machine prints out only true sentences, and the second machine prints out only false sentences. Let us call a sentence *provable* if and only if it is printable by the first machine and *refutable* if and only if it is printable by the second machine. Therefore, P can be read as "provable" and R can be read as "refutable."

4

Find a sentence which is false but not refutable.

5

There are sentences X and Y such that one of the two (we don't know which) must be either true but not provable or false but not refutable, and again we don't know which. This can be done in either of two ways, and I accordingly pose two problems:

(a) Find sentences X and Y such that X says that Y is provable and Y says that X is refutable. Then show that one of the sentences X or Y (we can't determine which) is either true and not provable or false but not refutable.

(b) Find sentences X and Y such that X says that Y is not provable and Y says that X is not refutable. Then show that for this X and Y, one of them (we can't determine which) is either true and not provable or false and not refutable.

6

Now let's try a quadruplicate! Find sentences X, Y, Z, and W such that X says that Y is provable, Y says that Z is refutable, Z says that W is refutable, and W says that X is not refutable. Show that one of these four sentences must be either true and not provable or false and not refutable (though there is no way to tell which of the four it is!).

McCULLOCH'S MACHINE AND GÖDEL'S THEOREM

The reader may have noticed certain similarities of some of the preceding problems to certain features of McCulloch's first machine. Indeed, this machine can be related to Gödel's theorem in the following manner:

7

Suppose we have a mathematical system with sentences certain ones of which are called *true* and certain of which are called *provable.* We assume the system is correct—every provable sentence is true. To each number N is assigned a sentence that we call *Sentence N*. Suppose the system satisfies the following two conditions:

Mc1: For any numbers X and Y, if X produces Y in McCulloch's first machine, then Sentence 8X is true if and only if Sentence Y is provable. (8X, remember, means 8 followed by X, not 8 times X.)

Mc2: For any number X, Sentence 9X is true if and only if Sentence X is not true.

Find a number N such that Sentence N is true but not provable in the system.

8

Suppose that in condition Mc1 of the last problem, we replace "McCulloch's first machine" by "McCulloch's third machine." *Now* find an N such that Sentence N is true but not provable!

9 • Paradoxical?

Let's return again to Problem 1, but with these differences: Instead of using the symbol "P," we will use "B" (for psychological reasons which will appear later). We define *sentence* as before, only this time using "B" in place of "P." Thus, our sentences are now B - X, NB - X, BA - X, and NBA - X. Sentences are, as before, classified into two groups—*true* sentences and *false* sentences—but we are not told which sentences are which. Now, instead of having a machine that prints out various sentences, we have a logician present who *believes* some of the sentences but not others. When we say that the logician *doesn't* believe a sentence, we don't mean that he *disbelieves* it; we merely mean that it is not the case that he believes it; in other words, he either believes it false or he has no opinion about it one way or the other. Now the symbol "B" stands for "believed by the logician," and so we are given that the following four conditions hold for any expression X:

B1: B – X is true if and only if the logician believes X.

B2: NB – X is true if and only if it is not the case that the logician believes X.

B3: BA – X is true if and only if the logician believes X – X.

B4: NBA – X is true if and only if it is not the case that the logician believes X – X.

Assuming that the logician is accurate—i.e., that he does not believe any false sentence—then we can, of course, find a sentence that is true but that the logician does not know to be true; namely, NBA – NBA (which says that the logician does not believe the associate of NBA, which is NBA – NBA).

Now comes something interesting. Suppose we are given the following facts about the logician:

Fact 1: The logician knows logic at least as well as you or I; in fact, we will assume that he is a perfect logician: given any premises, he can deduce all propositions that logically follow.

Fact 2: The logician knows that condition B1, B2, B3, and B4 all hold.

Fact 3: The logician is always accurate; he doesn't believe any false sentences.

Now, since the logician knows that conditions B1, B2, B3, and B4 all hold, and he can reason as well as you or I, what is to prevent *him* from going through the same reasoning process that we went through to prove that the sentence NBA – NBA must be true? It would appear that he can do this, and having done it, he will then believe the sentence

NBA – NBA. But the moment he believes it, the sentence will be falsified, since the sentence says that he doesn't believe it, which will make the logician inaccurate after all!

So, don't we get a paradox if we assume Facts 1, 2, and 3? The answer is that we don't; there is a deliberate flaw in my argument in this last paragraph! Can you find the flaw?

• SOLUTIONS •

1 • For any expression X, the sentence NPA – X says that the associate of X is not printable. In particular, NPA – NPA says that the associate of NPA is not printable. But the associate of NPA is the very sentence NPA – NPA! Hence, NPA – NPA asserts its own nonprintability; in other words, the sentence is true if and only if it is not printable. This means that either it is true and not printable or it is not true but printable. The latter cannot be, since the machine is accurate. Hence, it must be the former; the sentence is true but not printable by the machine.

2 • Let X be the sentence P – NPA – P – NPA and Y be the sentence NPA – P – NPA. The sentence X (which is P – Y) says that Y is printable. The sentence Y (crudely read as "not printable the associate of P – NPA") says that the associate of P – NPA is not printable. But the associate of P – NPA is X, so Y says that X is not printable. (Incidentally, there is another way of constructing such an X and Y: take X to be PA – NP – PA, and Y to be NP – PA – NP – PA.)

We thus have two sentences X and Y such that X says that Y is printable and Y says that X is not printable.

Now, suppose X were printable. Then it would be true, which would mean that Y is printable. Then Y would be true, which means that X is not printable. This is a contradiction, since X in this case would be both printable and not printable; hence X cannot be printable. Since X is not printable and Y says that X is not printable, then Y must be true. Therefore, we know:

(1) X is not printable;

(2) Y is true.

Now, X is either true or it isn't. If X is true, then, by (1), X is true but not printable. If X is false, then Y is not printable, since X says that Y *is* printable; and so in this case, Y is true—by (2)—and not printable. So either X is true and not printable or Y is true and not printable, but there is no way to tell which.

Discussion: The above situation is analogous to a knight-knave island on which there are two inhabitants X and Y, with X claiming that Y is an established knight and Y claiming that X is not an established knight. All that can be inferred is that at least one of them is an unestablished knight, but there is no way to tell which.

I deal with this situation in *What Is the Name of This Book?* in a section in the last chapter called "Doubly Gödelian Islands."

3 • We let Z = PA – P – NP – PA.

We let Y = NP – Z (which is NP – PA – P – NP – PA).

We let X = P – Y (which is P – NP – PA – P – NP – PA).

It is immediate that X says that Y is printable and Y says that Z is not printable. As for Z, Z says that the associate of P – NP – PA is printable, but the associate of P – NP – PA is P – NP – PA – P – NP – PA, which is X! So Z says that X is printable.

Thus X says Y is printable, Y says Z is not printable, and Z says X is printable. Now let us see what follows from this:

Suppose Z is printable. Then Z is true, which means that X

is printable, hence true, which means that Y is printable, hence true, which means that Z is not printable. So if Z is printable, it is not printable, which is a contradiction. Therefore, Z is not printable, and therefore Y is true. So we know:

(1) Z is not printable;

(2) Y is true.

Now, X is either true or false. Suppose X is true. If Z is false, then X is not printable, which means that X is true but not printable. If Z is true, then, since by (1) it is not printable, Z is true but not printable. So if X is true, then either X or Z is true but not printable. If X is false, then Y is not printable, hence Y is true—by (2)—and not printable.

In summary, if X is true, then at least one of the two sentences X and Z is true but not printable. If X is false, then it is Y that is true but not printable.

4 • Let S be the sentence RA – RA. It says that the associate of RA—which is S itself—is refutable; hence S is true if and only if S is refutable. Since S can't be true and refutable, it is therefore false but not refutable.

5 • (a) Take P – RA – P – RA for X and RA – P – RA for Y. Clearly, X says that Y is provable, and Y says that the associate of P – RA (which happens to be X) is refutable. So X says that Y is provable and Y says that X is refutable. (If we had taken PA – R – PA for X and R – PA – R – PA for Y, we would have had an alternative solution.)

Now, if Y is provable, then Y is true, which means that X is refutable, hence false, which means that Y is not provable. Thus we get a contradiction from the assumption that Y is provable, and therefore Y is not provable. Since Y is not provable, then X is false. And so we know:

(1) X is false;

(2) Y is not provable.

If Y is true, then Y is true and not provable. If Y is false, then X is not refutable (since Y says that X *is* refutable), and so in this case, X is false but not refutable. Therefore, either Y is true and not provable or X is false and not refutable.

(b) Take NP – NRA – NP – NRA for X, and NRA – NP – NRA for Y (or, alternatively, NPA – NR – NPA for X, and NR – NPA – NR – NPA for Y) and, as the reader can verify, X says that Y is not provable and Y says that X is not refutable. If X is refutable, X is false, Y is provable, Y is true, X is not refutable. Hence X is not refutable, and also Y is true. If X is false, X is false and not refutable. If X is true, Y is not provable; hence in this case Y would be true and not provable.

Discussion: Analogously, suppose we have inhabitants X and Y of a knight-knave island where X claims that Y is an established knight and Y claims that X is an established knave. All that can be deduced is that one of the two (we don't know which) must be either an unestablished knight or an unestablished knave. The same thing holds if X claims that Y is not an established knight and Y claims that X is not an established knave.

6 • Let W = NPA – P – R – R – NPA;
Z = R – W (which is R – NPA – P – R – R – NPA);
Y = R – Z (which is R – R – NPA – P – R – R – NPA);
X = P – Y (which is P – R – R – NPA – P – R – R – NPA).

X says that Y is provable, Y says that Z is refutable, Z says that W is refutable, and W says that X is not provable (W says that the associate of P – R – R – NPA, which is X, is not provable).

If W is refutable, W is false; hence X is provable, hence true; hence Y is provable, hence true; hence Z is refutable, hence false; hence W is not refutable. Therefore, it cannot be that W is refutable. So W is not refutable, and Z is therefore false.

Now, if W is false, then W is false but not refutable. Suppose W is true. Then X is not provable. If X is true, X is true and not provable. Suppose X is false. Then Y is not provable. If Y is true, then Y is true but not provable. Suppose Y is false. Then Z is not refutable; so in this case Z is false but not refutable.

This shows that either W is false and not refutable, or X is true and not provable, or Y is true and not provable, or Z is false and not refutable.

7 ◆ This situation is little more than a notational variant of this chapter's Problem 1!

We know that 32983 produces 9832983 (in McCulloch's first machine); hence by Mc1, Sentence 832983 is true if and only if Sentence 9832983 is provable. Also, by Mc2, Sentence 9832983 is true if and only if Sentence 832983 is not true; so, combining these last two facts, we see that Sentence 9832983 is true if and only if it is not provable. So the solution is 9832983.

If we compare this with Problem 1, we see that obviously, 9 plays the role of N, 8 plays the role of P, 3 plays the role of A, and 2 plays the role of the dash. Indeed, if we replace the symbols P,N,A,- by the respective numerals 8,9,3,2, the sentence NPA-NPA (which is the solution of Problem 1) becomes the number 9832983 (the solution of the present problem!).

8 ◆ To begin with, McCulloch's third machine also obeys McCulloch's Law—i.e., for any number A, there must be some X that produces AX. We prove this as follows: We know from Chapter 13 that there is a number H—namely,

5464—such that for any number X, H2X2 produces X2X2. (Recall that H2H2 then produces itself, but this is not relevant to the present problem.) Now, take any number A. Let X = H2AH2. Then X produces AH2AH2, which is AX. Thus X produces AX. And so for any number A, a number X that produces AX is 54642A54642.

We need an X that produces 98X: Suppose X does produce 98X. Then Sentence 8X is true if and only if Sentence 98X is provable (by Mc1); hence Sentence 98X is true if and only if Sentence 98X is not provable (by Mc2). Then Sentence 98X is true but not provable in the system (since the system is correct).

Now, from the last paragraph, taking 98 for A, we see that an X which produces 98X is 546429854642. Hence Sentence 98546429854642 is true but not provable in the system.

9 • I told you that the logician was accurate, but I never told you that he *knew* he was accurate! If he knew he was accurate, then the situation *would* lead to a paradox! Therefore, what properly follows from Facts 1, 2, and 3 is not a contradiction but simply that the logician, though accurate, cannot know that he is accurate.

This situation is not totally unrelated to another theorem of Gödel's, known as *Gödel's Second Incompleteness Theorem,* which (roughly speaking) states that for systems with a sufficiently rich structure (and this includes the systems treated in Gödel's original paper), if the system is consistent, then it cannot prove its own consistency. This is a profound matter, which I plan to discuss further in a sequel to this book.

◆ 17 ◆
Mortal and Immortal Numbers

It had been some time since Craig had last seen McCulloch or Fergusson, when he met the two of them quite unexpectedly one late afternoon and the three happily went off to dine together.

"You know," said McCulloch, after the meal, "there is one problem that has baffled me for quite a while."

"And what is that?" asked Fergusson.

"Well," replied McCulloch, "I have studied several machines, and with each one I run into the same problem: In each of the machines, certain numbers are acceptable and others are not. Now, suppose I feed an acceptable number X into the machine. The number Y that X produces is either unacceptable or acceptable. If Y is unacceptable, the process terminates; if Y is acceptable, I feed Y back into the machine to see what number Z is produced by Y. If Z is unacceptable, then the process terminates; if Z is acceptable, I feed it back into the machine, and so the process continues for at least one more cycle. I repeat this over and over again, and there are two possibilities: one, I eventually get an unacceptable number; two, the process goes on forever. If the former, then I call X a *mortal* number, with respect to the machine in question, and if the latter, I call X an *immortal* number. Of course, a given number might be mortal for one machine and immortal for another machine."

"Let's consider your first machine," said Craig. "I can think of plenty of mortal numbers, but can you give me an example of an immortal number?"

"Obviously, 323," replied McCulloch. "323 produces itself, so if I put 323 into the machine, out comes 323. I put 323 back, and again out comes 323. So, in this case, the process clearly never terminates."

"Oh, of course!" laughed Craig. "Are there other immortal numbers?"

1

"Well," replied McCulloch, "what would you say about the number 3223? Is it mortal or not?"

2

"What about the number 32223?" asked Fergusson. "Is it mortal or immortal for your first machine?"

McCulloch thought about this for a bit. "Oh, that's not too difficult to settle," he replied. "I think you might enjoy trying your hand at it."

3

"You might also try the number 3232," said McCulloch. "Is this number mortal or immortal?"

4

"What about the number 32323?" asked Craig. "Mortal or immortal?"

5

"These are all fine questions," said McCulloch, "but I haven't yet come to the main problem. A friend of mine has constructed a rather elaborate number machine, which he claims can do anything that any machine can do; he calls it a *universal* machine. Now, there are several numbers of which neither he nor I can tell whether they are mortal or immortal, and I would like to devise some purely mechanical test to determine which numbers are which, but so far I have not succeeded. Specifically, I am trying to find a number H such that for any acceptable number X, if X is immortal then HX is mortal, and if X is mortal then HX is immortal. If I could find such a number H, then I could decide for any acceptable number X whether X is mortal or immortal."

"How would finding such an H enable you to do that?" asked Craig.

"If I had such a number H," replied McCulloch, "I would first build a duplicate of my friend's machine. Then, given any acceptable number X, I would feed X into one of the machines, and at the same time my friend would feed HX into the other machine. One and only one of the processes would terminate; if the first process terminated, then I would know that X is mortal; if the second process terminated, then I would know that X is immortal."

"You wouldn't actually have to build a second machine,"

said Fergusson. "You could alternate the stages of the two processes."

"True," replied McCulloch, "but this is all hypothetical, since I have not been able to find such a number H. Perhaps this machine *can't* solve its own mortality problem—that is, perhaps there *is* no such number H. Then again, maybe I just haven't been clever enough to find it. This is the problem about which I would like to consult you gentlemen."

"Well," replied Fergusson, "we must know the rules of this machine. What are they?"

"There are twenty-five rules," began McCulloch. "The first two are the same as those of my first machine."

"Just a moment," said Fergusson. "Are you saying that your friend's machine obeys your Rules 1 and 2?"

"Yes," replied McCulloch.

"Well, that settles the matter!" replied Fergusson. "No machine obeying Rules 1 and 2 can possibly solve its own mortality problem!"

"How can you have determined that so quickly?" asked Craig.

"Oh, this isn't new to me," replied Fergusson. "A similar problem came up in my own work some time ago."

How *did* Fergusson know that no machine obeying Rules 1 and 2 can solve its own mortality problem?

• SOLUTIONS •

1 • We recall that 3223 produces 23223, and of course 23223 produces 3223. So, we have the two numbers 3223 and 23223, each of which produces the other. So, they are both immortal: Put one of them into the machine and the other

comes out; put the second back into the machine and the first comes out. The process clearly never terminates.

2 ◆ For any two numbers X and Y, let us say that X *leads to* Y if either X produces Y, or X produces some number that produces Y, or X produces some number that produces some number that produces Y, or X produces some number that produces some number that . . . that produces some number that produces Y. Stated otherwise, if starting the process with X we get Y at some state or other, then we will say that X leads to Y. As an example, 22222278 leads to 78—after six stages, in fact. More generally, if T is any string of 2's, then for any number X, TX leads to X.

Now, 32223 does not produce itself, but it does lead to itself, because it produces 2232223, which in turn produces 232223, which in turn produces 32223. Since 32223 leads to itself, it must be immortal.

The reader might note the following more general fact: For *any* number T that consists entirely of 2's, the number 3T3 must lead to itself, and hence must be immortal.

3 ◆ The only way I know to solve this problem is to show the more general fact that for any T consisting entirely of 2's, the number 3T32 is immortal, and that therefore the specific instance 3232 is immortal. And this illustrates a still more general principle which will also be used in the solution of the next problem.

Suppose we have a class of numbers (whether the class is finite or infinite makes no difference), and the class is such that every member of the class leads to some member of the class (either itself or some other member). Then every member of the class must be immortal.

To apply this principle to the problem at hand, let us consider the class of all numbers of the form 3T32, where T is a string of 2's. We will show that 3T32 must lead to another number of this class.

Let us first consider the number in question, 3232. It produces 32232, which is a member of this class. What about 32232? It produces 2322232, which in turn produces 322232, which is a member of this class. What about 322232? It produces 223222232, which produces 23222232, which produces 3222232, so we are again back in this class. More generally, for any string T of 2's, 32T32 produces T322T32, which leads to 322T32, which is again a member of this class. So all members of this class are immortal.

4 • The number 32323 produces 3232323, which produces 32323232323, which produces 3232323232323232323. The pattern should be obvious: any number consisting of 32 repeated some number of times and then followed by 3 produces another number of this form (a longer one, in fact), and so all such numbers are immortal.

5 • We first observe the following fact: Suppose X and Y are numbers such that X produces Y. Then if Y is mortal, X must also be mortal, because if Y leads to an unacceptable number Z in n stages, then X will lead to Z in $n + 1$ stages. Also, if Y is immortal, it never leads to any unacceptable number; hence X cannot lead to an unacceptable number, since the only way X can lead to a number is via Y. So, if X produces Y, then the mortality of X is the same as the mortality of Y (i.e., they are either both mortal or both immortal).

Now, consider any machine that obeys at least Rules 1 and 2 (and possibly others). Take any number H. We know that by Rules 1 and 2, there must be a number X that produces HX (indeed, we recall that H32H3 is such a number). Since X produces HX, then the numbers X and HX are either both mortal or both immortal (as we showed in the last paragraph). So there cannot be any number H such that for *every* X, one of the numbers H or HX is mortal and the other immortal, because for the particular number X = H32H3, it is not the case that one of the numbers X or HX is mortal and the other immortal. Therefore, no machine obeying Rules 1 and 2 can solve its own mortality problem.

We might remark that the same goes for any machine obeying Rules 1 and 4, or indeed for any machine obeying McCulloch's Law. (This whole problem, incidentally, is closely related to a famous *halting problem* of Turing machines, whose solution is also negative.)

◆ 18 ◆

The Machine That Never Got Built

Shortly after the last episode, Craig was sitting quietly in his study one early afternoon. There was a timid knock at the door.

"Pray come in, Mrs. Hoffman," said Craig to his landlady.

"There is a wild, eccentric-looking gentleman to see you, sir," said Mrs. Hoffman. "He claims to be on the verge of the greatest mathematical discovery of all time! He says it would interest you *enormously* and insists on seeing you immediately. What should I do?"

"Well," replied Craig judiciously, "you might as well send him up; I have about half an hour to spare."

A few seconds later, the door of Craig's study burst open, and a distracted, frenzied inventor (for an inventor he was) practically flew into the room, flung his briefcase on a nearby sofa, threw up his hands, danced wildly around the room, shouting, "Eureka! Eureka! I'm about to find it! It will make me the greatest mathematician of all time! Why, the names of Euclid, Archimedes, Gauss will pale into insignificance! The names of Newton, Lobochevski, Bolyai, Riemann . . ."

"Now, now!" interrupted Craig, in a quiet but firm voice, "just what is it that you have found?"

"I haven't exactly *found* it yet," replied the stranger, in a somewhat more subdued tone. "But I'm *about* to find it, and when I do, I'll be the greatest mathematician who ever lived! Why, the names of Galois, Cauchy, Dirichlet, Cantor . . ."

"Enough!" interrupted Craig. "Please tell me just what it is you are trying to find."

"*Trying* to find?" said the stranger, with a somewhat hurt expression. "Why, I tell you, I've *almost* found it! A universal machine which can solve *all* mathematical problems! Why, with this machine, I'll be omniscient! I'll be able to . . ."

"Ah, Leibniz's dream!" said Craig. "Leibniz also had such a dream, but I doubt that the dream is realizable."

"Leibniz!" said the stranger, contemptuously. "Leibniz! He just didn't know how to go about it! But I practically *have* such a machine! All I need is a couple of details—but here, let me give you a concrete idea of what I am after:

"I am looking for a machine M," explained the stranger (whose name, it turned out, was Walton), "with certain properties: To begin with, you put a natural number x into the machine and after that a natural number y; then the machine goes into operation and out comes a natural number which we'll call $M(x,y)$. So $M(x,y)$ is the output of M when the input is x as the first number and y as the second number."

"I'm with you so far," said Craig.

"Now, then," continued Walton, "I shall use the word *number* to mean *positive integers,* since the positive integers are the only numbers I will be considering. As you may know, two natural numbers are said to have the same parity if they are either both even or both odd, and they are said to be of different parities if one of them is even and the other one is odd.

"For every x, let $x^{\#}$ be the number $M(x,x)$. Now, here are the three properties I want my machine to have:

"*Property 1:* For every number a, I want there to be a number b such that for every number x, $M(x,b)$ has the same parity as $M(x^{\#},a)$.

"*Property 2:* For every number b, I want there to be a number a such that for every x, $M(x,a)$ has a different parity than $M(x,b)$.

"*Property 3:* I want there to be a number h such that for every x, $M(x,h)$ has the same parity as x.

"These are the three properties I want my machine to have," concluded Walton.

Inspector Craig thought about this for some time.

"Then what is your problem?" he finally asked.

"Alas," replied Walton. "I have built a machine having Properties 1 and 2, and another having Properties 1 and 3, and a third having Properties 2 and 3. All these machines work perfectly—indeed, I have complete plans for them in my briefcase over there—but when I try to put the three properties together in one machine, something goes wrong!"

"Just what is it that goes wrong?" asked Craig.

"Why, the machine doesn't work at all!" cried Walton, with an air of desperation. "When I put in a pair (x,y) of numbers, instead of getting an output, I get a strange buzzing sound—something like a short circuit! Do you have any idea why that is?"

"Well, well!" said Craig. "This is something I'll have to think about. I must be off now on a case, but if you'll leave your card—or, if you haven't one, your name and address—I will let you know if I can arrive at any solution."

Several days later, Inspector Craig wrote a letter to Walton that began as follows:

My dear Mr. Walton:

I thank you for your visit and for calling my attention to the machine that you are trying to build. To be perfectly honest, I cannot quite see how, even if you actually constructed such a machine, it would be capable of solving *all* mathematical problems, but doubtless you understand this matter better than I. More to the point, however, I must tell you that your project is much like trying to build a perpetual-motion machine: it simply cannot be done! Indeed, the situation here is even worse—for a perpetual-motion machine, though not possible in this physical world, is not *logically* impossible; whereas such a machine as you propose is not merely physically impossible, but logically impossible, since the three properties you mention conceal a logical contradiction.

Craig's letter then went on to explain just why the existence of such a machine is a logical impossibility. Can you see why?

It will be helpful to break up the solution into three steps:

(1) Show that for any machine having Property 1, for any number a, there must be at least one number x such that $M(x,a)$ has the same parity as x.

(2) Show that for any machine having Properties 1 and 2, for any number b, there is some number x such that $M(x,b)$ has a different parity from x.

(3) No machine can have Properties 1, 2, and 3 combined.

◆ SOLUTION ◆

(a) Consider a machine with Property 1. Take any number a. By Property 1 there is a number b such that for every x, $M(x,b)$ has the same parity as $M(x^{\#},a)$. In particular, with b taken for x, $M(b,b)$ has the same parity as $M(b^{\#},a)$. However

$M(b,b)$ is the number $b^{\#}$, so $b^{\#}$ has the same parity as $M(b^{\#},a)$. So, letting x be the number $b^{\#}$, we see that $M(x,a)$ has the same parity as x.

(b) Consider now any machine having properties 1 and 2. Take any number b. By Property 2 there is some a such that for every x, $M(x,a)$ has different parity than $M(x,b)$. And, by Property 1, there is at least one x such that $M(x,a)$ has the same parity as x, as we proved in (a) above. Such an x must then have different parity from $M(x,b)$—because it has the same parity as $M(x,a)$, which has different parity from $M(x,b)$.

(c) Again consider a machine having properties 1 and 2. Take any number h. According to (b) above, (reading "h" for "b") there is at least one x such that $M(x,h)$ has different parity than x. Therefore, it cannot be that for *all* numbers x, $M(x,h)$ has the same parity as x; in other words, Property 3 cannot hold. Thus, Properties 1, 2, and 3 are "incompossible" (to use Ambrose Bierce's lovely term!).

Note: The impossibility of Walton's machine is closely related to Tarski's theorem (Chapter 15), and it is not difficult to prove the theorem and the machine's impossibility by a common argument.

◆ 19 ◆
Leibniz's Dream

Fergusson (as well as Walton, in his own peculiar way) was attempting something which, if successful, would fulfil one of Leibniz's most fervent dreams: Leibniz envisioned the possibility of a calculating machine that could solve all mathematical problems—and all philosophical ones as well! Leaving aside the philosophical problems, it appears that for even the mathematical ones, Leibniz's dream is not feasible. This follows from the results of Gödel, Rosser, Church, Kleene, Turing, and Post, to whose work we now turn.

There is a type of computing machine whose function is to calculate a mathematical operation on the positive integers. For such a machine, you feed in a number x (the input) and out comes a number y (the output). For example, you can easily design a machine (not a very interesting one, to be sure!) such that whenever a number x is fed in, out comes the number $x + 1$. Such a machine may be said to *compute* the operation of adding 1. Or we might have a machine that computes an operation, on two numbers, such as addition. For such a machine, you first feed in a number x, then a number y; then you press a button and, after a while, out comes the number $x + y$. (There is a technical name for such machines, of course—I believe they are called *adding machines!*)

There is another type of machine that might be called a *generating* or *enumerating* machine, which will play a more fundamental role in the approach we will take here (it follows the theories of Post). Such a machine has no inputs; it is programmed to generate a set of positive integers. For example, we might have one machine to generate the set of even numbers, another to generate the set of odd numbers, another to generate the set of prime numbers, and so forth. A typical program for a machine to generate the even numbers might run something like the following.

We give the machine two instructions: (1) that it print out the number 2; (2) that if ever it prints out a number n it may also print out $n + 2$. (You also give auxiliary rules that *systematize* following the instructions, so that anything the machine *can* do, it eventually *will* do.) Such a machine, obeying Instruction (1), will sooner or later print out 2, and having printed 2, it will sooner or later, by Instruction (2), print out 4, and having printed 4 will sooner or later print 6, again by Instruction (2), then 8, then 10, and so forth. This machine, then, will generate the set of even numbers. (Without further instructions, it could never come out with 1, 3, 5, or any of the odd numbers.) To program a machine to generate the set of odd numbers, of course, we need merely change the first instruction to: "Print out 1." Sometimes two or more machines are coupled so that the output of one machine can be used by one of the other machines. For example, suppose we have two machines, A and B, and program them as follows: To A we give two instructions: "(1) Print out 1; (2) if ever Machine B prints out n, then print out $n + 1$." To Machine B we give only one instruction: "(1) If ever Machine A prints out n, then print out $n + 1$." What set will A generate, and what set will B generate? The answer is that A will generate the set of odd numbers and B will generate the set of even numbers.

Now, a program for a generating machine, instead of being given in English, is coded into a positive integer (in the form of a string of digits) and matters can be arranged so that every positive integer is the number of some program. We let M_n be the machine whose program has code number n. We now think of all generating machines as listed in the infinite sequence $M_1, M_2, \ldots, M_n, \ldots$ (M_1 is the machine whose program number is 1, M_2 the machine whose program number is 2, and so forth.)

For any number set A (set of positive integers, that is) and any machine M, we shall say that M *generates* A or, alternatively, that M *enumerates* A, if every number in A is eventually printed out by M; but no number outside A ever gets printed out by M. We shall say that A is *effectively enumerable* (another technical term is *recursively enumerable*) if there is at least one machine M_i that enumerates A. We shall say that A is *solvable* (another technical term is *recursive*) if there is one machine M_i that enumerates A and another machine M_j that enumerates the set of all numbers that are not in A. Thus, A is solvable if and only if both A and its complement $\bar{A}$ are effectively enumerable.

Suppose A *is* solvable, and we are given a machine M_i that generates A and a machine M_j that generates the complement of A. Then we have an effective procedure to determine whether any number n lies inside A or outside A: Suppose, for example, we wish to know whether the number 10 is in A or not. We set both machines, M_i and M_j, going simultaneously and wait. If 10 lies inside A, then sooner or later M_i will print out 10, and we will know that 10 belongs to A. If 10 lies outside A, then sooner or later machine M_j will print out

10, and we will know that 10 doesn't belong to A. So, eventually and inevitably, we will know whether 10 belong to A or whether it doesn't. (Of course, we have no idea in advance of how long we will have to wait; all we know is that in *some* finite time we will know the answer.)

Now, suppose a set A is effectively enumerable but not solvable. Then we have a machine M_i that generates A, but we have no machine to generate the complement of A. Suppose again we would like to know whether a given number—say, 10—is or is not in A. The best we can do in this case is to set the machine M_i going and hope for the best! We now have only a 50 percent chance of ever learning the answer. If 10 *is* in A, then sooner or later we will know it, because sooner or later M_i will print out 10. If 10 is not in A, however, M_i will never print out 10, but no matter how long we wait, we will have no assurance that M_i might not print out 10 at some later time. So if 10 is in A, we will sooner or later know that it is, but if 10 isn't in A, then at no time can we definitely know that it isn't (at least by observing only the machine M_i). We might aptly call such a set A *semisolvable.*

The first important feature of these generating machines is that it is possible to design a so-called *universal* machine U whose function is to observe systematically the behavior of all the machines M_1, M_2, . . ., M_n, . . ., and whenever a machine M_x prints out a number y, U is to report the fact. How does it make this report? By printing out a number: for any x and y we again let $x*y$ be the number, which consists of a string of 1's of length x followed by a string of 0's of length y. Our principal instruction to U is: "Whenver M_x prints out y, then print out $x*y$."

Suppose, for example, that M_a is programmed to generate the set of odd numbers, and M_b is programmed to generate the set of even numbers. Then U will print out all the numbers $a*1$, $a*3$, $a*5$, $a*7$, etc., and also all the numbers $b*2$, $b*4$, $b*6$, $b*8$, etc., but U will never print out $a*4$ (since M_a never prints out 4), nor $b*3$ (since M_b never prints out 3).

Now, the machine U itself has a program, hence is one of the programmable machines $M_1, M_2, \ldots, M_n, \ldots$ Thus, there is a number k such that M_k is the very machine U! (In a more complete technical account of the matter, I could tell you what the number k is.)

We might note that this universal machine M_k observes and reports its own behavior as well as that of all the other machines. So whenever M_k prints out a number n, it must also print out $k*n$, hence also $k*(k*n)$, hence also $k*[k*(k*n)]$, and so forth.

A second important feature of these machines is that for any machine M_a we can program a machine M_b to print out those and only those numbers x such that M_a prints out $x*x$. (M_b, so to speak, "keeps watch" on M_a and is instructed to print out x whenever M_a prints $x*x$.) It is possible to code programs in such a way that for each a, $2a$ is such a number b; that is, for every a, M_{2a} prints out those and only those numbers x such that M_a prints out $x*x$. We will assume this done, and so let us record two basic facts that will be used in what follows:

Fact 1: The universal machine U prints out those and only those numbers $x*y$ such that M_x prints out y.

Fact 2: For every number a, the machine M_{2a} prints out those and only those numbers x such that M_a prints out $x*x$.

We now come to the central issue: Any formal mathematical problem can be translated into a question of whether a machine M_a does or does not print out a number b. That is, given any formal axiom system, one can assign Gödel numbers to all the sentences of the system and find a number a

such that the machine M_a prints out the Gödel numbers of the provable sentences of the system, and no other numbers. So to find out whether a given sentence is or is not provable in the system, we take its Gödel number b and ask whether the machine M_a does or does not print out b. Thus, if we had some effective method of deciding which machines print out which numbers, we could then effectively decide which sentences are provable in which axiom systems. This would constitute a realization of Leibniz's dream. Moreover, the question of which machines print out which numbers can be reduced to the question of which numbers are printed out by the universal machine U, because the question of whether or not machine M_a prints out b is equivalent to the question of whether or not U prints out the number $a*b$. Therefore, a complete knowledge of U would entail a complete knowledge of all the machines, and hence of all mathematical systems. Conversely, any question of whether a given machine prints out a given number can be reduced to a question of whether a certain sentence is provable in a certain mathematical system, and so a complete knowledge of all formal mathematical systems would imply a complete knowledge of the universal machine.

The key question, then, is this: Let V be the set of numbers printed out by the universal machine U (this set V is sometimes called the *universal set*). Is this set V solvable or not? If it is, then Leibniz's dream is realized; if it isn't, then Leibniz's dream cannot ever be realized. Since V is effectively enumerable (it is generated by the machine U), the question boils down to whether or not there is some machine M_a that prints out the *complement* $\bar{V}$ of V; that is, is there a machine M_a which prints out those and only those numbers that U does *not* print? This question can be completely answered just on the basis of the given conditions Facts 1 and 2 above.

Theorem L: The set $\bar{V}$ is not effectively enumerable: Given any machine M_a, either there is some number in $\bar{V}$ that M_a fails to print, or $\bar{M}$ prints at least one number that is in V rather than $\bar{V}$.

Can the reader see how to prove Theorem L? To take a special case, suppose the claim were made that the machine M_5 enumerated $\bar{V}$. To disprove this claim, it would suffice to exhibit a number n and show that either n is in $\bar{V}$ and M_5 fails to print n or n is in V and M_5 prints n. Can you find such a number n?

I shall give the solution now rather than at the end of the chapter. The solution is really Gödel's argument over again:

Take any number a. By Fact 2, for every number x, M_a prints $x{*}x$ if and only if M_{2a} prints x. But, also by Fact 1, M_{2a} prints x if and only if the universal machine U prints $2a{*}x$, or, what is the same thing, if and only if $2a{*}x$ is in the set V. Therefore, M_a prints $x{*}x$ if and only if $2a{*}x$ is in V. In particular (taking $2a$ for x), M_a prints the number $2a{*}2a$ if and only if $2a{*}2a$ is in V. So either: (1) M_a prints $2a{*}2a$ and $2a{*}2a$ is in V; (2) M_a doesn't print $2a{*}2a$ and $2a{*}2a$ is in $\bar{V}$. If (1) holds, then M_a prints out the number $2a{*}2a$, which is not in $\bar{V}$ but in V; this means that M_a does not generate the set $\bar{V}$, because it prints out at least one number ($2a{*}2a$), which is not in $\bar{V}$. If (2) holds, then again M_a fails to generate the set $\bar{V}$, because the number $2a{*}2a$ is in $\bar{V}$ but fails to get printed by M_a. So in neither case does M_a generate the set $\bar{V}$. Since no machine can generate $\bar{V}$, the set $\bar{V}$ is not effectively enumerable.

Of course, for the special case $a = 5$, the number n is $10{*}10$.

Now, what does all this mean with respect to Leibniz's dream? Strictly speaking, one cannot prove or disprove the feasibility of Leibniz's hope, because it was not stated in an exact form. Indeed, no precise notion of a "calculating machine" or "generating machine" existed in Leibniz's day; these notions have been rigorously defined only in this century. They have been defined in many different ways (by Gödel, Herbrand, Kleene, Church, Turing, Post, Smullyan, Markov, and many others), but all these definitions have been shown to be equivalent. If by "solvable" is meant solvable according to any of these equivalent definitions, then Leibniz's dream is not feasible, because the simple fact is that the machines can be numbered in such a way that Facts 1 and 2 do hold. Then, by Theorem L, the set V generated by the universal machine is not solvable; it is only semisolvable. Therefore, there is no purely "mechanical" procedure for finding out which sentences are provable in which axiom systems and which ones are not. Thus, any attempt to invent a clever "mechanism" that will solve all mathematical problems for us is simply doomed to failure.

In the prophetic words of the logician Emil Post (1944), this means that mathematical thinking is, and must remain, essentially creative. Or, in the witty comment of the mathematician Paul Rosenbloom, it means that man can never eliminate the necessity of using his own intelligence, regardless of how cleverly he tries.

About the Author

Curiously enough, I have lived four different lives—as a mathematician, musician, magician, and author of essays and puzzle books. I was born in 1919 in Far Rockaway, New York. As a child, I was equally interested in music and science. In high school I fell in love with mathematics, and was torn between becoming a mathematician or a concert pianist. My first teaching position was at Roosevelt College in Chicago, where I taught piano. At about that time I unfortunately developed tendonitis in my right arm forcing me to abandon piano performances as my primary career. As a result of this I turned my attention to mathematics. I had learned most of this on my own, with very little formal education at the time. I then took a few advanced courses at the University of Chicago, and supported myself at the time as a professional magician!

Strangely enough, before I had a college degree, or even a high school diploma, I received an appointment as a mathematics instructor at Dartmouth College on the basis of some papers I had written on mathematical logic. After teaching at Dartmouth, the University of Chicago gave me a Bachelor of Arts degree, based partly on courses I had never taken, but had successfully taught, such as calculus. I then went to Princeton University for my Ph.D. in mathematics in 1959. I subsequently taught at Princeton, NYU, Belfor Graduate School, Lehman College and Graduate Center, and my last teaching position was as a distinguished ranks professor at Indiana University. I have published forty research papers in mathematical logic and twenty-two books. This coming year [2009] I am coming out with four more books.

Despite the trouble with my right arm, I have been able to give concerts, and am still musically active. I am a working member of the Piano Society. I continue to write puzzle books, but these are more than mere puzzle books—it is through recreational logic puzzles that I introduce the general reader to deep results in mathematics and logic!

RAYMOND M. SMULLYAN